Adhesives and Adhesive Joints in Industry Applications

Edited by Anna Rudawska

Published in London, United Kingdom

IntechOpen

Supporting open minds since 2005

Adhesives and Adhesive Joints in Industry Applications
http://dx.doi.org/10.5772/intechopen.77485
Edited by Anna Rudawska

Contributors
Petr Belov, Sergey Lurie, Golovina Nataly, Chulsoo Woo, Jaber Khanjani, David Michael Strumpf, Chunfu Chen, Syam Kumar Chokka, Beera Satish Ben, Sai Srinadh K. V

First published in London, United Kingdom, 2019 by IntechOpen
IntechOpen is the global imprint of INTECHOPEN LIMITED, registered in England and Wales, registration number: 11086078, The Shard, 25th floor, 32 London Bridge Street
London, SE19SG – United Kingdom
Printed in Croatia

British Library Cataloguing-in-Publication Data
A catalogue record for this book is available from the British Library

Additional hard and PDF copies can be obtained from orders@intechopen.com

Adhesives and Adhesive Joints in Industry Applications
Edited by Anna Rudawska
p. cm.
Print ISBN 978-1-78984-898-4
Online ISBN 978-1-78984-899-1
eBook (PDF) ISBN 978-1-83962-258-8

For EU product safety concerns:
IN TECH d.o.o., Prolaz Marije Krucifikse Kozulić 3, 51000 Rijeka, Croatia, info@intechopen.com or visit our website at intechopen.com.

Meet the editor

Anna Rudawska, PhD, DSc, Eng, Associate Professor, is head of the Technological Processes and Systems Design Chair and since September 1, 2012 she has been Vice Dean of the Faculty of Mechanical Engineering, Lublin University of Technology, Poland. She is the author or coauthor of 230 scientific publications and several patents in the field of adhesives. She is a member of a number of international and national organizations, and a reviewer and member of editorial boards of international journals. Her research activities are focused on the issues of analysis of the influence of technological and structural factors on polymer and metal adhesive joint strength, design of bonding technology, adhesive properties research, as well as issues related to receiving the appropriate adhesive properties to increase adhesive joint strength.

Contents

Preface

This book discusses applications of adhesives and adhesive joints in different branches of industry. The properties of adhesives and adhesive joints, and also the requirements of mechanical properties and chemical and environmental resistance of adhesives and adhesive joints, are very important because proper strength, durability, and time of use are all factors that are dependent on the type of industry.

The book is divided into two parts: *Adhesive and adhesive joint properties* and *Applications of adhesives and adhesive joints*. The aim of such a presentation is to show the characteristics and applications of adhesives and adhesive joints in different branches of industry. Recent applications in industry require increased cost saving and more effective and reliable assembly of various parts of different products such as machines, planes, cars, buses, structures, ships, etc. Some chapters provide an insight into the solutions of different functional bonding and sealing technologies available to manufacturers who have hitherto used conventional methods in the hope of helping them to choose and apply alternative solutions.

The aim of this book is to present information on the types of adhesives and adhesives joints and also their characteristics used in different branches of industry, because this information should enable scientists, engineers, and designers to acquire knowledge of adhesives and adhesive joints, which could be helpful in selecting the right type of adhesive and adhesive joint for particular industries.

I would like to express my gratitude to all of the contributors for their high-quality manuscripts. I hope the open access format of this book will help both researches and engineers and that they will benefit from its content.

Anna Rudawska
Lublin University of Technology,
Faculty of Mechanical Engineering,
Department of Production Engineering,
Lublin, Poland

Section 1

Adhesive and Adhesive Joints Properties

Chapter 1

Classifying the Existing Continuum Theories of Ideal-Surface Adhesion

Belov Petr Anatolyevich, Lurie Sergey Albertovich and Golovina Natalya Yakovlevna

Abstract

The chapter classifies the existing continuum theories of ideal-surface adhesion within the gradient theory of adhesion. Ideal surface herein means a defect-free surface, the deformed state of which is entirely defined by the displacement vector and its first (distortion) derivatives as well as its second (curvature) derivatives. Ideal surfaces have such kinematic variables as noncombined deformations and rotations. The classification is based on a formal quadratic form of potential surface energy, which comprises contracting the first-rank tensors (adhesive-force theory), second-rank tensors (adhesive-stress theory), and third-rank tensors (theory of adhesive couple stresses). To interpret the physical sense of the summands in the quadratic form of the potential-energy surface density, this research uses a rather common method of dividing the elastic solid into an internal solid plus a surface layer (adhesive, contact, boundary, or inter-phase layer). The formal structure of the adhesion-energy surface density is compared to the structure of the thickness-averaged potential energy of a selected 3D layer. The chapter establishes the most general structure of adhesive-moduli tensors for the surfaces of classical elastic solids. The adhesive modules specific to the surfaces of a solid in gradient elasticity theories are identified.

Keywords: continuum adhesion theories, adhesive moduli, adhesive interaction, scale effects, nonclassical physical parameters

1. Introduction

Recent investigations of adhesive properties of surfaces and interfaces in deformable solids, in the mechanics of heterogeneous structures, and in the mechanics of composites are developed in various publications and analyzed in detail [1–10]. The first adhesion continuum theories were developed in the framework of the classical theory of elasticity [1, 2, 11–14].

The theory of Gurtin-Murdoch [11], which has become classical, was called as the theory of elasticity of surfaces. A generalization of this theory is proposed in the paper [15].

The adhesion theories listed above determined the adhesion properties of ideal (defect-free) surfaces. Further generalization of the theory of adhesion on the surface of defective media is given in [16–20].

Belov and Lurie [19, 20] formulated a model in which the adhesive properties were attributed to the newly formed surface connected with a field of defects. A variational model that takes into account the adhesive interactions of perfect (not damaged) surfaces, surfaces damaged by defects, and their interaction was presented. The surface of the defective medium can be represented as a perfect surface and a defective surface. Each of them has its own adhesive properties, as well as the properties of interaction with each other.

Adhesive interactions between the inclusion and the matrix in fine composites [18] are of great interest, as they directly affect not only the stiffness, but also the strength properties of composites.

So, as the adhesive properties of the surface are determined not only by the tangential derivatives of the displacements but also the normal derivatives, the boundary value problems for a classical body can be redefined due to the presence of adhesive interactions proportional to the normal derivatives of the displacements. On the other hand, in gradient theories, in nonclassical boundary conditions, the inclusion of adhesive interactions gives new effects. In other words, the surface of the body consists of the surface of a classical body and the surface of a "gradient" body. They have different adhesion properties and can even interact with each other as well as a defect-free (ideal) surface and a surface damaged by defects.

Similarly, to the gradient theories of elasticity, which contain the quadratic form of the second derivatives of displacements in potential energy, there are adhesion theories that also take into account the quadratic form of the second derivatives of displacements in the potential adhesion energy. A gradient theory of second order, which can be considered as a generalization of the theory of Steigmann and Ogden [1, 2], is described in Belov and Lurie [20].

The purpose of this chapter is the sequential analysis of variational formulations of the theories of adhesive interactions and the classification of adhesion models by the degree of accuracy of accounted scale effects. Classification of theories of adhesion and gradient theories of elasticity in terms accounting for scale effects was proposed in the work [21].

We have the following statement regarding the general structure of the adhesion elastic moduli for the classical linearly elastic body [15, 17, 18].

In line with this, the Lagrangian L of the model is written as:

$$L = A - \iiint U_V dV - \oiint U_F dF \tag{1}$$

Here, $A = \iiint P_i^V u_i dV + \oiint P_i^F u_i dF$ is the work of the volumetric forces P_i^V and surface forces P_i^F during the displacements u_i; U_V is the potential-energy density; and U_F is the potential-energy surface density.

The difference between the potential surface energies of two solids in contact in each contact spot is what determines their adhesive interaction. This is why adhesion theories can be classified on the basis of the potential-energy surface density inherent in an isolated solid.

A general expression for the potential-energy surface density U_F for an ideal surface is written as:

$$2U_F = A_{ij}u_i u_j + A_{ijmn}u_{i,j}u_{m,n} + A_{ijkmnl}u_{i,jk}u_{m,nl} + \dots \tag{2}$$

where u_i, $u_{i,j}$, and $u_{i,jk}$ are the displacement vector, its first derivatives, and second derivatives, respectively; A_{ij}, A_{ijmn}, A_{ijkmnl} are tensors of the rank-specific adhesive moduli, which are transversely isotropic to the unit normal vector to the surface n_i. According to Green's formulas, each summand in (2) corresponds to a specific set of adhesive-force factors: adhesive forces a_i, adhesive stresses a_{ij}, or adhesive couple stresses a_{ijk}, etc.

$$\begin{aligned} a_i &= \frac{\partial U_F}{\partial u_i} = A_{ij} u_j \\ a_{ij} &= \frac{\partial U_F}{\partial u_{i,j}} = A_{ijmn} u_{m,n} \\ a_{ijk} &= \frac{\partial U_F}{\partial u_{i,jk}} = A_{ijkmnl} u_{m,nl} \end{aligned} \tag{3}$$

Accordingly, adhesive-moduli tensors are structured as follows [20]:

$$A_{ij} = a_n n_i n_j + a_s \delta^*_{ij} \tag{4}$$

$$\begin{aligned} A_{ijmn} = {} & \lambda^F \delta^*_{ij} \delta^*_{mn} + \mu^F \left(\delta^*_{im} \delta^*_{jn} + \delta^*_{in} \delta^*_{jm} \right) + \alpha^F \left(n_i n_n \delta^*_{jm} + n_m n_j \delta^*_{in} \right) \\ & + \beta^F \left(n_i n_j \delta^*_{mn} + n_m n_n \delta^*_{ij} \right) + \delta^F n_i n_m \delta^*_{jn} + B^F \delta^*_{im} n_j n_n + A^F n_i n_j n_m n_n \end{aligned} \tag{5}$$

$$\begin{aligned} A_{ijkmnl} = {} & A_1 \Big(\delta^*_{ij} \delta^*_{km} \delta^*_{nl} + \delta^*_{mn} \delta^*_{li} \delta^*_{jk} + \delta^*_{ij} \delta^*_{kn} \delta^*_{ml} + \delta^*_{mn} \delta^*_{lj} \delta^*_{ik} \\ & + \delta^*_{ij} \delta^*_{kl} \delta^*_{mn} + \delta^*_{ik} \delta^*_{jm} \delta^*_{nl} + \delta^*_{ml} \delta^*_{ni} \delta^*_{jk} + \delta^*_{in} \delta^*_{km} \delta^*_{jl} + \delta^*_{mj} \delta^*_{li} \delta^*_{nk} \\ & + \delta^*_{in} \delta^*_{lk} \delta^*_{jm} + \delta^*_{ik} \delta^*_{jn} \delta^*_{ml} + + \delta^*_{il} \delta^*_{km} \delta^*_{nj} + \delta^*_{im} \delta^*_{kj} \delta^*_{nl} + \delta^*_{im} \delta^*_{lj} \delta^*_{nk} \\ & + \delta^*_{im} \delta^*_{nj} \delta^*_{kl} \Big) + A_2 \left(n_i n_m \delta^*_{kj} \delta^*_{nl} + n_i n_m \delta^*_{lj} \delta^*_{nk} + n_i n_m \delta^*_{nj} \delta^*_{kl} \right) \\ & + A_3 \Big(n_i n_j \delta^*_{km} \delta^*_{nl} + n_m n_n \delta^*_{li} \delta^*_{jk} + n_i n_j \delta^*_{kn} \delta^*_{ml} + n_m n_n \delta^*_{lj} \delta^*_{ik} \\ & + n_i n_j \delta^*_{kl} \delta^*_{mn} + n_m n_n \delta^*_{lk} \delta^*_{ij} \Big) + A_4 \Big(n_i n_n \delta^*_{km} \delta^*_{jl} + n_m n_j \delta^*_{li} \delta^*_{nk} \\ & + n_i n_n \delta^*_{ml} \delta^*_{jk} + n_m n_j \delta^*_{ik} \delta^*_{nl} + n_i n_n \delta^*_{lk} \delta^*_{jm} + n_m n_j \delta^*_{kl} \delta^*_{ni} \Big) \\ & + A_5 \left(n_j n_n \delta^*_{ik} \delta^*_{ml} + n_j n_n \delta^*_{il} \delta^*_{km} \right) + A_6 n_j n_n \delta^*_{im} \delta^*_{kl} + A_7 \delta^*_{kl} n_i n_j n_m n_n \end{aligned} \tag{6}$$

Here, n_i is the unit normal vector to the surface and $\delta^*_{ij} = \left(\delta_{ij} - n_i n_j \right)$ is the planar Kronecker tensor $\left(\delta^*_{ij} n_j = \delta^*_{ij} n_i = 0, \delta^*_{ij} \delta^*_{ij} = 2 \right)$.

2. Theory of adhesion with adhesive forces

The first summand in Eq. (2) identifies the contribution made by the "spring" adhesion model [22]. The model derives its name from the specific nature of the corresponding adhesive forces a_i. In the spring theory of adhesion, adhesive forces are proportional to displacement, which enables comparing them to the response of the Winkler foundations from the classical theory of elasticity. In the spring model, provided that the surface properties are isotropic, there are two adhesive parameters per (4): the stiffness of the normal spring a_n and that of the tangential spring a_s.

There is an approach based on comparing the adhesive properties to the properties of a fictitious finite-thickness surface layer; this approach can be reduced to the spring theory. The algorithm of reducing a 3D surface layer to a spring model consists in finding its thickness such that the deformations in the real contact surface and in the surface layer are equivalent [23]. The disadvantage here is that the algorithm cannot explain the adhesive properties of 2D structures such as graphene or single-wall nanotubes, since such structures feature no thickness and therefore no surface layer.

The algorithm can be demonstrated by a simple example. Let an elastic solid be presented as an internal solid plus a surface layer, or Skin, the thickness h whereof is so small compared to the total size of the solid that the deformed state of this layer can be deemed homogeneous. Then, the distribution of displacements in this layer can be deemed linear across its thickness, which is equivalent to Timoshenko's kinematic hypotheses [24]:

$$\begin{cases} u = u_0 + u_1\left(\dfrac{z}{h} - \dfrac{1}{2}\right) \\ v = v_0 + v_1\left(\dfrac{z}{h} - \dfrac{1}{2}\right) \\ w = w_0 + w_1\left(\dfrac{z}{h} - \dfrac{1}{2}\right) \end{cases} \tag{7}$$

Note that $w_1 = w(h) - w(0)$ is the relative normal displacement of the corresponding points on the opposite sides of the "surface" layer; $u_1 = u(h) - u(0)$ and $v_1 = v(h) - v(0)$ are the projections of the tangential relative displacements; and $\sqrt{u_1 u_1 + v_1 v_1}$ is the magnitude of tangential relative displacements. The classical-medium Lagrangian expression can be rewritten to obtain the following equalities:

$$\begin{aligned} L &= A - \iiint_V U_V dV = A - \iiint_{V-Skin} U_V dV - \iiint_{Skin} U_V dV = \\ &= A - \iiint_{V-Skin} U_V dV - \oiint \left(\int_0^h U_V dz \right) dF = \\ &= A - \iiint_{V-Skin} U_V dV - \oiint U_F dF \end{aligned} \tag{8}$$

For a classical elastic solid, the volumetric portion of the "surface-layer" potential energy is written as:

$$\begin{aligned} \int_0^h U_V dz = \frac{1}{2}\{ & 2\mu\left[\left(u_{0,x}u_{0,x} + v_{0,y}v_{0,y}\right)h + \left(u_{1,x}u_{1,x} + v_{1,y}v_{1,y}\right)h/12\right] \\ & + \lambda\left[\left(u_{0,x} + v_{0,y}\right)^2 h + \left(u_{1,x} + v_{1,y}\right)^2 h/12\right] + \mu\left[\left(u_{0,y} + v_{0,x}\right)^2 h \right.\\ & \left. + \left(u_{1,y} + v_{1,x}\right)^2 h/12\right] + \mu\left[\left(w_{0,x}w_{0,x} + w_{0,y}w_{0,y}\right)h \right.\\ & \left. + \left(w_{1,x}w_{1,x} + w_{1,y}w_{1,y}\right)h/12\right] + 2\mu\left(w_{0,x}u_1 + w_{0,y}v_1\right) \\ & + 2\lambda\left(u_{0,x} + v_{0,y}\right)w_1 + \mu\left(u_1u_1 + v_1v_1\right)/h + (2\mu + \lambda)w_1w_1/h\} \end{aligned} \tag{9}$$

The "adhesive properties" of this layer depend on many factors, including the relative displacement of corresponding points on its fronts. Note that the summands in the first four lines of the expression (9) are proportional to h^{+1}, the summands in the fifth line are proportional to h^0, while the summands in the last line are proportional to h^{-1}. Provided a sufficiently thin layer, these summands must make the greatest contribution to the potential-energy expression as long as all the layer deformations belong to the same order. These components are the energy of springs resisting the relative normal and tangential displacement of the layer front points; they determine the spring stiffness:

$$a_n = (2\mu + \lambda)/h \tag{10}$$

$$a_s = \mu/h \tag{11}$$

The stiffness values are found via Young's modulus, the surface-layer shear, and the unknown thickness h.

Equations (10) and (11) indicate that in order for the spring stiffness values to be independent, it is necessary to abandon the condition that the "surface layer" modules must be the same as the modules inside a solid. Indeed, excluding the "surface-layer" thickness from (10), (11) means that the springs are proportional stiffness-wise:

$$a_s = a_n\mu/(2\mu + \lambda) \tag{12}$$

this contradicts the initial assumption that the adhesive properties of an isolated-solid surface (1) do not depend on the internal mechanical properties of the same solid; it also contradicts the empirical data, according to which one and the same solid may have different adhesive properties depending on what chemical or physicochemical method was used to activate its surface [25].

Thus, expression (1) *is entirely different from the adhesive surface layer model* (8), (9) *even in the context of the simplest spring model*. Potential adhesion energy in (1) will have the following structure for the spring theory of adhesion:

$$2U_F = A_{ij}u_iu_j \tag{13}$$

while the tensor of second-rank adhesive moduli A_{ij} is structured as:

$$A_{ij} = a_n n_i n_j + a_s \delta^*_{ij} \tag{14}$$

3. Adhesion theories with adhesive stresses

As is outlined above, *expressions* (8) and (9) *are not directly applicable to the potential energy of adhesion*; they only enable structuring the potential energy of adhesion in (2) and finding some physical analogies and interpretations. In particular, the first summand in (2) can be interpreted as the potential energy of springs and the stiffness whereof has the same physical dimensionality [Pa/m] as shown in the ratios (10) and (11). In doing so, it is stated that the surface-layer deformations belong to the same order. Let us now assume the opposite. Assume that the traverse deformations are negligible compared to longitudinal deformations. Then the summands of the last and last-but-one lines in the expression (9) can be ignored. Let us study such expression (9) for the potential "surface-layer" energy further to find analogy to the second summand in (2).

4. Gurtin-Murdoch theory (surface elasticity theory)

The first, the third, the fifth, the tent, and the twelfth summands in (9) determine the potential tension-compression and shear energy in their corresponding planes of the surface layer [11]. If Timoshenko's static hypothesis (the noncompressibility hypothesis) is introduced to the kinematic hypotheses (7), then the kinematic variable w_i can be expressed in terms of the changing area $(u_{0,x} + v_{0,y})$; then, the summands above are reduced to being written in terms of the potential energy of the layer during tension or compression (Line 1) or in the case of shear (Line 2):

$$\int_0^h U_V dz|_{M-G} = \frac{1}{2}h\left\{4\mu\frac{(\mu+\lambda)}{(2\mu+\lambda)}u_{0,x}u_{0,x} + 2\frac{2\mu\lambda}{(2\mu+\lambda)}u_{0,x}v_{0,y} + 4\mu\frac{(\mu+\lambda)}{(2\mu+\lambda)}v_{0,y}v_{0,y} + \mu(u_{0,y}+v_{0,x})^2\right\}$$

Accordingly, the potential adhesion energy in (1) will have the following structure for the Gurtin-Murdoch theory of adhesion [12]:

$$2U_F = A_{ijmn}u_{i,j}u_{m,n} \tag{15}$$

while the tensor of fourth-rank adhesive moduli A_{ijmn} is structured as:

$$A_{ijmn} = \lambda^F\delta^*_{ij}\delta^*_{mn} + \mu^F\left(\delta^*_{im}\delta^*_{jn} + \delta^*_{in}\delta^*_{jm}\right) \tag{16}$$

This interprets the first to modules of the fourth-rank tensor in (5). Apparently, the adhesive moduli λ^F and μ^F are adhesive analogs to the Lame coefficients for the planar stress-state problem in the classical theory of elasticity. What must not be forgotten is that adhesive moduli have a different dimensionality: [Pa/m] rather than [Pa].

5. Belov-Lurie theory (theory of ideal adhesion)

The Belov-Lurie theory of ideal adhesion [15] can be formulated by adding a summand similar to the seventh summand $\mu h(w_{0,x}w_{0,x} + w_{0,y}w_{0,y})$ in (9) to the summands that define the Gurtin-Murdoch theory in (2). The potential energy of adhesion per (1) is structured as follows for this adhesion theory:

$$2U_F = A_{ijmn}u_{i,j}u_{m,n} \tag{17}$$

while the tensor of fourth-rank adhesive moduli A_{ijmn} is structured as:

$$A_{ijmn} = \lambda^F\delta^*_{ij}\delta^*_{mn} + \mu^F\left(\delta^*_{im}\delta^*_{jn} + \delta^*_{in}\delta^*_{jm}\right) + \delta^F n_i n_m \delta^*_{jn} \tag{18}$$

Paper [15] shows that the adhesive modulus δ^F is the Laplace capillary pressure constant if the expressions (17) and (18) apply to liquids. This modulus is critical for the adhesive properties of surfaces in solids, as it takes into account the adhesion effects of the same order as in the Gurtin-Murdoch theory. The latter is incomplete in this respect. Paper [26] compares the Gurtin-Murdoch theory against the Belov-Lurie theory in terms of modeling the adhesive interactions between the edges of cracks that differ in mode. It shows that the statements of boundary problems for mode II and mode III cracks will coincide for both theories. However, for mode I cracks, the ideal-adhesion theory takes into account the adhesive properties of the crack edges, while the classical statement or Gurtin-Murdoch statement does not.

6. Theory of adhesion surfaces of classical media

It shows that the adhesive properties of surfaces of a solid are closely related to the models of the solid itself. Indeed, if the internal properties of a solid are defined by the classical theory of elasticity, i.e., $U_V = U_V(u_{i,j}) = C_{ijmn}u_{i,j}u_{m,n}$, then (1) means that:

$$\delta L = \delta A - \iiint \sigma_{ij}\delta u_{i,j}dV - \oiint (a_i\delta u_i + a_{ij}\delta u_{i,j} + a_{ijk}\delta u_{i,jk})dF =$$

$$= \iiint (\sigma_{ij,j} + P_i^V)\delta u_i dV + \oiint \left\{ \left[P_i^F - \sigma_{ij}n_j - a_i \right.\right.$$

$$\left.\left. + \left(a_{ij} - a_{ijk,q}\delta^*_{kq}\right)_{,p}\delta^*_{pj}\right]\delta u_i - \left(a_{ij} - a_{ijk,q}\delta^*_{kq}\right)n_j\delta \dot{u}_i \right\}dF = 0 \quad (19)$$

For simplicity, consider an edge-less solid, i.e., a solid that has a smooth rather than piecewise-smooth surface. Analysis of the variational Eq. (19) shows that the boundary problem for the classical theory of elasticity is redefined, as a correctly formulated system will have three boundary conditions in each nonsingular surface point. At the same time, the variational Eq. (19) provides six boundary conditions in each non-singular surface point. To remove this contradiction, it must be required that the force factors always equal zero on the surface of classical elastic solids when varying the $\delta\dot{u}_i$. This requirement is equivalent to the conditions of adhesion-moduli tensors:

$$A_{ijmn}n_j = \alpha^F n_m\delta^*_{in} + \beta^F n_i\delta^*_{mn} + B^F\delta^*_{im}n_n + A^F n_i n_m n_n \equiv 0 \quad (20)$$

$$A_{ijkmnl}n_j = A_3\left(n_i\delta^*_{km}\delta^*_{nl} + n_i\delta^*_{kn}\delta^*_{ml} + n_i\delta^*_{kl}\delta^*_{mn}\right)$$

$$+A_4\left(n_m\delta^*_{li}\delta^*_{nk} + n_m\delta^*_{ik}\delta^*_{nl} + n_m\delta^*_{kl}\delta^*_{ni}\right) + A_5\left(n_n\delta^*_{ik}\delta^*_{ml} + n_n\delta^*_{il}\delta^*_{km}\right)$$

$$+A_6 n_n\delta^*_{im}\delta^*_{kl} + A_7\delta^*_{kl}n_i n_m n_n \equiv 0 \quad (21)$$

Thus, the adhesion theories for the surfaces of classical elastic media are limited by the following structure for the potential adhesion energy:

$$2U_F = A_{ij}u_i u_j + A_{ijmn}u_{i,j}u_{m,n} + A_{ijkmnl}u_{i,jk}u_{m,nlj} \quad (22)$$

In the context of (4) and (18), the tensors of adhesive moduli will have the following structure that is maximally general for a surface that confines a classical medium:

$$A_{ij} = a_n n_i n_j + a_s\delta^*_{ij} \quad (23)$$

$$A_{ijmn} = \lambda^F\delta^*_{ij}\delta^*_{mn} + \mu^F\left(\delta^*_{im}\delta^*_{jn} + \delta^*_{in}\delta^*_{jm}\right) + \delta^F n_i n_m\delta^*_{jn} \quad (24)$$

$$A_{ijmnkl} = A_1\left(\delta^*_{ij}\delta^*_{km}\delta^*_{nl} + \delta^*_{mn}\delta^*_{li}\delta^*_{jk} + \delta^*_{ij}\delta^*_{kn}\delta^*_{ml} + \delta^*_{mn}\delta^*_{lj}\delta^*_{ik}\right.$$

$$+\delta^*_{ij}\delta^*_{kl}\delta^*_{mn} + \delta^*_{ik}\delta^*_{jm}\delta^*_{nl} + \delta^*_{ml}\delta^*_{ni}\delta^*_{jk} + \delta^*_{in}\delta^*_{km}\delta^*_{jl} + \delta^*_{mj}\delta^*_{li}\delta^*_{nk}$$

$$+\delta^*_{in}\delta^*_{lk}\delta^*_{jm} + \delta^*_{ik}\delta^*_{jn}\delta^*_{ml} + \delta^*_{il}\delta^*_{km}\delta^*_{nj} + \delta^*_{im}\delta^*_{kj}\delta^*_{nl} + \delta^*_{im}\delta^*_{lj}\delta^*_{nk}$$

$$\left.+\delta^*_{im}\delta^*_{nj}\delta^*_{kl}\right) + A_2 n_i n_m\left(\delta^*_{kj}\delta^*_{nl} + \delta^*_{lj}\delta^*_{nk} + \delta^*_{nj}\delta^*_{kl}\right) \quad (25)$$

This seems to be the first time to state the above.

7. Steigmann-Ogden theory (instantaneous adhesion theory)

Transforming the surface-layer potential energy expression (9) according to Kirchhoff's rather than Timoshenko's hypotheses, the energy is written as the potential plate-bending energy:

$$\int_0^h U_V dz|_{S-0} = \frac{1}{2}\{Dw_{0,xx}w_{0,xx} + 2vDw_{0,xx}w_{0,yy} + Dw_{0,yy}w_{0,yy} + 2(1-v)Dw_{0,xy}w_{0,xy}\} \tag{26}$$

In this case, the summands in (27) have a structure that corresponds to the particular case (26), where $A_1 = 0$. In this case, the potential adhesion energy depends only on the second tangent derivatives of the surface deflections [22]. In the general case, it must also depend on the second tangent derivatives of the displacement projections tangential to the surface of a solid. Thus, the Steigmann-Ogden theory [2] is a special case of the gradient adhesion theory [27], which is defined by the structure of the sixth-rank adhesive-moduli tensor (6) for Toupin's gradient media, or by the tensor of moduli (26) for classical elastic solids.

8. Gradient-media surface adhesion theories

Let the internal properties of a solid be determined by Toupin's gradient theory, i.e.,

$$U_V = U_V(u_{i,j}, u_{i,jk}) = (C_{ijmn}u_{i,j}u_{m,n} + C_{ijkmnl}u_{i,jk}u_{m,nl})/2$$

Then (1) means that:

$$\begin{aligned}\delta L = \delta A - \iiint(\sigma_{ij}\delta u_{i,j} + \sigma_{ijk}\delta u_{i,jk})dV - \oiint(a_i\delta u_i + a_{ij}\delta u_{i,j} \\ + a_{ijk}\delta u_{i,jk})dF = \iiint(\sigma_{ij,j} - \sigma_{ijk,kj} + P_i^V)\delta u_i dV \\ + \oiint\Big\{\Big[P_i^F - \sigma_{ij}n_j + \sigma_{ijk,k}n_j + (\sigma_{ijk}n_k)_{,p}\delta^*_{pj} \\ -\Big(a_i - \big(a_{ij} - a_{ijk,q}\delta^*_{kq}\big)_{,p}\delta^*_{pj}\Big)\Big]\delta u_i - \Big[\sigma_{ijk}n_jn_k \\ +\big(a_{ij} - a_{ijk,q}\delta^*_{kq}\big)n_j\Big]\delta\dot{u}_i\Big\}dF = 0\end{aligned} \tag{27}$$

As in the case of deriving the variational Eq. (19), for simplicity, consider an edgeless solid, i.e., a solid that has a smooth rather than piecewise-smooth surface. Analysis of the variational Eq. (28) shows that the boundary problem for Toupin's gradient theory is correct, as the formulated system will have six pairs of alternative boundary conditions in each non-singular surface point. Note that the requirements (20) that define the structure of adhesive tensors for classical-solid surfaces also specify a group of adhesive moduli that only manifest on the surfaces of Toupin's solids. It can therefore be stated that the adhesive moduli $\alpha^F, \beta^F, B^F, A^F$ and A_3, A_4, A_5, A_6, A_7 are specific only to the gradient theories of elasticity as generalized by Toupin's theory. This also seems to be the first time to state the above.

9. Graphene-type surface adhesion theories

Consider the ratio (1) in its extreme case, in which the volumetric portion of the potential energy of the solid is negligible compared to the surface energy:

$$L = A - \oiint U_F dF \tag{28}$$

The potential-energy surface density structure is the same as in (2). Note that the normal derivative to the surface is not defined for kinematic variables in 2D structures. This is why the structure (29) can be the most general structure for the adhesive-moduli tensors [30].

Paper [28] considers the properties of graphene-like structures without the "spring" portion of potential energy. It has been found that graphene features the bending properties of Timoshenko's plates, while its cylindrical stiffness is determined by the adhesive modulus A_2 whereas the shear stiffness is determined by the adhesion modulus δ^F. Therefore, the mechanical tension-compression and planar shear properties are determined by the remaining moduli λ^F, μ^F and A_1.

Paper [29–31] considers the properties of single-walled nanotubes (SWNT), the potential energy of which is written specifically as (2).

$$2U_F = A_{ijmn} u_{i,j} u_{m,n} + A_{ijkmnl} u_{i,jk} u_{m,nl}$$

The resultant finding is that the mechanical properties of single-walled nanotubes are determined by a nonclassical modification of the cylindrical-shell theory equations.

10. Conclusion

This chapter presents an attempt to classify the existing continuum theories of ideal-surface adhesion within the gradient theory of adhesion [20, 28, 31]. The classification is based on a formal quadratic form of potential surface energy (2), which comprises contracting the first-rank tensors (adhesive-force theory), second-rank tensors (adhesive-stress theory), and third-rank tensors (theory of adhesive couple stresses). To interpret the physical sense of the summands in the quadratic form of the potential-energy surface density, this research uses a rather common method of dividing the elastic solid into an internal solid plus a surface layer (6) (adhesive, contact, boundary, or inter-phase layer). The formal structure of the adhesion-energy surface density (2) is compared to the structure of the potential energy averaged over the thickness of the selected subsurface 3D layer (7). There has been found the most general structure of adhesive-moduli tensors for the surfaces of classical elastic solids (20). The chapter identifies the adhesive moduli specific to the surfaces of solids in gradient theories of elasticity (18).

Acknowledgements

This work was carried out particularly with support from the Russian Government Foundation of Institute of Applied Mechanics of RAS, Project AAAA-A19-119012290177-0.

Author details

Belov Petr Anatolyevich[1], Lurie Sergey Albertovich[1]
and Golovina Natalya Yakovlevna[2*]

1 Institute of Applied Mechanics of RAS, Moscow, Russia

2 Industrial University of Tyumen, Surgut, Russia

*Address all correspondence to: ntgolovina@rambler.ru

References

[1] Steigmann DJ, Ogden RW. Plane deformations of elastic solids with intrinsic boundary elasticity. Proceedings of the Royal Society of London. 1997;**453**:853-877. DOI: 10.1098/rspa.1997.0047

[2] Steigmann DJ, Ogden RW. Elastic surface-substrate interactions. Proceedings of the Royal Society of London. 1999;**455**:437-474. DOI: 10.1098/rspa.1999.0320

[3] Ostoja-Starzewski M. Lattice models in micromechanics. Applied Mechanics Reviews. 2002;**55**:35-59

[4] Duan HL, Wang J, Huang ZP, Karihaloo BL. Size-dependent effective elastic constants of solids containing nanoinhomogeneities with interface stress. Journal of the Mechanics and Physics of Solids. 2005;**53**:1574-1596

[5] Duan HL, Wang J, Karihaloo BL. Theory of elasticity at the nanoscale. In: Aref H, van der Giessen E, editors. Advances in Applied Mechanics. Vol. 42. Amsterdam: Elsevier; 2009. pp. 1-68. DOI: 10.1016/S0065-2156(08)00001-X

[6] Zhu HX, Wang JX, Karihaloo BL. Effects of surface and initial stresses on the bending stiffness of trilayer plates and nanofilms. Journal of Mechanics of Materials and Structures. 2009;**4**: 589-604

[7] Altenbach H, Eremeyev VA, Lebedev LP. On the spectrum and stiffness of an elastic body with surface stresses. ZAMM. 2011;**91**(9):699-710. DOI: 10.1002/zamm.201000214

[8] Povstenko Y. Mathematical modeling of phenomena caused by surface stresses in solids. In: Altenbach H, Morozov NF, editors. Surface Effects in Solid Mechanics. Vol. 30. Heidelberg: Springer; 2013. pp. 135-153. DOI: 10.1007/978-3-642-35783-1_11

[9] Huang Z, Wang J. Micromechanics of nanocomposites with interface energy effect. In: Li S, Gao XL, editors. Handbook on Micromechanics and Nanomechanics. Vol. 42. Stanford: Pan Stanford Publishing; 2013. pp. 303-348

[10] Eremeyev VA. On effective properties of materials at the nano- and microscales considering surface effects. Acta Mechanica. 2016;**227**:29-42. DOI: 10.1007/s00707-015-1427-y

[11] Gurtin ME, Murdoch AI. Addenda to our paper a continuum theory of elastic material surfaces. Archive for Rational Mechanics and Analysis. 1975; **59**(4):389-390. DOI: 10.1007/BF00250426

[12] Gurtin ME, Murdoch AI. A continuum theory of elastics material surface. Archive for Rational Mechanics and Analysis. 1975;**57**(4):291-323. DOI: 10.1007/BF00261375

[13] Eremeyev VA, Altenbach H, Morozov NF. The influence of surface tension on the effective stiffness of nanosize plates. Doklady Akademii Nauk. 2009;**424**(5):618-621. DOI: 10.1134/S102833580902013X

[14] Kim CI, Schiavone P, Ru CQ. Effect of surface elasticity on an interface crack in plane deformations. Proceedings of the Royal Society A. 2011;**467**(2136):3530-3549. DOI: 10.1098/rspa.2011.0311

[15] Belov PA, Lurie SA. The theory of ideal adhesive interactions. Mechanics of Composite Materials and Structures. 2007;**3**(V14):519-536

[16] Mindlin RD. Second gradient of strain and surface-tension in linear

elasticity. International Journal of Solids and Structures. 1965;**1**(4):417-438. DOI: 10.1016/0020-7683(65)90006-5

[17] Lurie SA, Belov PA, Solyaev YO. Adhesion interactions in continuum mechanics. Mathematical Modeling of Systems and Processes. 2008;**16**:76-86

[18] Lurie SA, Volkov-Bogorodsky DB, Zubov VI, Tuchkova NP. Advanced theoretical and numerical multiscale modeling of cohesion/adhesion interactions in continuum mechanics and its applications for filled nanocomposites. Computational Materials Science. 2009;**45**(3):709-714. DOI: 10.1016/j.commatsci.2008.08.010

[19] Belov PA, Lurie SA. Continual theory of adhesion interactions of damaged media. Journal of Composite Materials. 2009;**15**(4):610-629

[20] Belov PA, Lurie SA. Mathematical theory of damaged media. Gradient theory of elasticity. Formulations hierarchy comparative analysis. In: Analysis and Applications. Palmarium Academic Publishing; 2014. 336 p. ISBN: 978-3-639-73761-5

[21] Lurie S, Belov P, Altenbach H. Classification of gradient adhesion theories across length scale. In: Altenbach H, Forest S, editors. Generalized Continua as Models for Classical and Advanced Materials. Vol. 42. Cham: Springer; 2016. pp. 261-277. DOI: 10.1007/978-3-319-31721-2_13

[22] Goryacheva IG, Makhovskaya YY. Modeling friction at different scale levels. Solid Mechanics. 2010;**3**:100-110

[23] Turusov RA, Manevich LI. Contact layer method in adhesion mechanics. Adhesives, Sealants, Technologies. 2009;**6**:2-11

[24] Vasiliev VV. Mechanics of Structures Made of Composite Materials. Moscow: Mashinostroenie; 1988. p. 264 p. ISBN: 5-217-00038-4

[25] Nelyub VA, Belov PA. Methods to improve the resistance of carbon fiber to delamination by controlling the adhesion of the fiber to the matrix. Material Science News. Science and Technology. 2013;**5**:1-11

[26] Belov PA, Nelub VA. Choice of the theory of adhesive interactions for interlayered cracks modelling in composites. Mechanics of Composite Materials. 2016;**51**(6):745-750. DOI: 10.1007/s11029-016-9545-8

[27] Belov PA. Theory of media with conserved dislocations: A generalization of the Mindlin model. Composites and Nanostructures. 2011;**3**(1):24-38

[28] Belov PA. Mechanical properties of graphene within the framework of gradient theory of adhesion. Journal of Civil Engineering and Architecture. 2014;**8**(6):693-698. DOI: 10.17265/1934-7359/2014.06.003

[29] Belov PA, Lurie SA, Eremeev VN. Continuum Model of the SWNT Based on the Generalized Theory of Adhesion Interactions. In: XII International Conference on Nanostructured Materials (NANO 2014); 13–18 July 2014; Moscow

[30] Belov PA. Mechanical properties of SWNT within the framework of gradient theory of adhesion. Journal of Nanoscience and Nanoengineering. 2015;**1**(4):218-224. ISSN: 2471-8378

[31] Belov PA, Lurie SA. Mechanical properties of SWNT within the framework of the theory of ideal adhesion. Journal of Nanoscience and Nanoengineering. 2017;**3**(2):6-10

Chapter 2

Evolving the Future of Smart Adhesives: Adhesives Utilizing Layers, Composites and Substrates with Smart Materials

David Strumpf

Abstract

The focus of this chapter is to present emerging technologies within the category of smart adhesives which have the potential to grow into a significant industry segment of adhesives, coatings, layers and hybrid chemistry solutions. We will examine the history, evolution and potential direction of materials and techniques that could yield functional and practical smart adhesives of the future.

Keywords: smart adhesive, damping, resonance, nanotube, thermal response

1. Introduction

1.1 Smart adhesives: evolution

As adhesive innovations continue to advance industries can exploit attributes of discrete material discoveries, hybrids and applications in order to create new engineered materials, composites, layers and coatings to strive toward adhesive compounds that are so amazing they appear to be "smart" [1]. It is important to note that these advancements, enhancements and innovations will simply represent the 'State of the Art' for a moment in time. We live in an exciting time as advancements are occurring at exponential rates which are bringing many opportunities for rapid innovation through cross-industry solutions. This means that science and technology are rapidly reaching intersections that allow applications to be realized as scalable solutions in size and ease of deployment. Scale can be thought of as macro-, micro- and nano-scaled. Macro may be a component that you hold in your hand such as a washer, gasket or o-ring. Micro may be a mixture of components that are apparent to the naked eye and Nano may be a molecular array, bonding, layering or complementary mating that is smaller than can be seen but can be quite powerful in application value.

1.2 Background adhesive review

Ever since their development, adhesives have been used in every industry for everything from constructing containers to adhering shingles to a roof, and nearly everything in between. Adhesives provide several advantages over other binding

IntechOpen

techniques such as sewing, mechanical fastening, etc., including the ability to bind together dissimilar materials, make design choices that would otherwise be unachievable, and more efficiently distribute stresses across a joint, to name a few. However, adhesives also suffer from several disadvantages. Adhesives may experience decreased stability at less than ideal conditions (i.e., at high or low temperatures). Further, the larger the objects, the more difficult it becomes to adhere the objects together if the bonding surface area is small. Finally, where a high degree of adhesion is desirable so that the materials do not become separated, it may be difficult to also provide a sufficient degree of flexibility to allow the materials to expand and contract due to changes in the environment surrounding the adhered objects. As a result, the adhesive, or even the adhered object itself, may malfunction.

Properties of desired adhesive considerations include:

- Bonding
- Thermal
- Durometer
- Viscosity
- Resonance
- Bonding
- Resilience

2. Smart materials: background overview

Smart materials are becoming an important part of our future, and in many cases they are right under (or above) our noses today [1]. Every-day examples range from your child's forehead thermometer to your sunglass frames.

Children's forehead film thermometers are inexpensive and convenient tools for measuring approximate temperatures based on the surface reaction of materials in a layered film. But what makes it "smart"? The particles that are layered and positioned in the thermometer film are specially selected inks and polymer particles that change their reflective properties based on the surface temperature. The resulting effect creates the appearance of intelligence by providing a viewpoint of perspective to the user. The particles are individually performing a specific localized reaction to a stimulus that is fundamental to the particles themselves. In this case the attribute of the particles is that they change their optically reflective properties based on temperature. This gives the user the apparent viewpoint that the film appears to become "smart" by providing a reading of temperature in color and numerical information.

Another smart material example that we use every day is built into most of our sunglass frames through the use of shape memory alloy material. One of the most amazing and spectacular metal alloys in use today is nickel-titanium [2] (also known as Nitinol). One of the properties of this material is the ability to program a shape of the metallic structure under extreme temperatures. This seemingly magical property is commonly called Shape-Memory. The resulting shape can be recalled through material memory by reheating the object to a nominal temperature, which will cause realignment of the particles to the original programmed shape. Other uses for Nitinol alloy is a form of wire called muscle wire. The wire shape can be

DOI: http://dx.doi.org/10.5772/intechopen.85785

formed in such a way that it effectively contracts its length like a muscle when heated. Another principle of the material is that it is electrically conductive.

Therefore the programmable stimulus can be in the form of electrical energy and perform motion based on a computer signal. This is very important for the advancements in robotics and industrial automation.

What is a smart material?

So what is a smart material? The generally accepted definition is, "A material that exhibits one or more properties that can be significantly changed in a controlled fashion by external stimuli" [3]. You could therefore generalize this definition to say that everything and every one of us is a smart material. But, in the context of material science today we focus our definitions to the nonobvious and spectacular.

I want to make this point because the perception of spectacular and common knowledge is an ever-changing spectrum. In the not-too-distant future we will find potential to control molecules in ways that were not previously possible. An example of this is in the ever-expanding discoveries based on DNA and protein modeling of matter [4].

We are all examples of the spectacular nature of smart materials that have been programmed by DNA coding becoming organisms and humans. We are also continually programmed by external stimuli reacting and responding to ever-changing situations. The fact that we gain a greater understanding than we previously had does not make these now-known facts less spectacular.

For example, the most basic discovery of moving particles based on temperature was shown clearly through the invention of the Galileo Thermometer in the 1600s [5]. This was an amazing accomplishment of a physical set of attributes based on object density and fluid-object displacement.

The temperature can be observed by the user in a self-contained temperature acquisition and display device. A truly amazing invention for its time and still respected today. However, we do not think of this as spectacular in today's high-tech world.

Many scientific discoveries become obsolete over time and industries lose interest in the scientific fundamentals of the inventions and discoveries that once were thought to be magical and spectacular. Another example of a scientific principle that is still widely used today but is becoming obsolete is the principle of bimetal displacement based on temperature. This technology is one of the world's simplest thermostat inventions (**Figure 1**). The movement of the two different metals occurs due to each metal having different attributes based on temperature. The net effect is a predictable movement of the combined structure of bimetal commonly known as a bimetal strip [6]. You can imagine that the movement of the heated and cooled bimetal shape could be connected to an ON/OFF switch where the switch is pushed upward to turn ON the air conditioner and pushed downward to turn OFF the air conditioner. This is the basic structure of almost all mechanical thermostats made over the last hundred years. One interesting note related to this "magical" phenomenon is that we do not think of this solution as a smart material in today's modern world.

So, where is the world of discovery heading and what should we look for as significant smart materials for the future?

Areas of accelerated development in smart materials:

- Polymers
- Ceramics
- Carbon forms
- Mixed-mode hybrids.

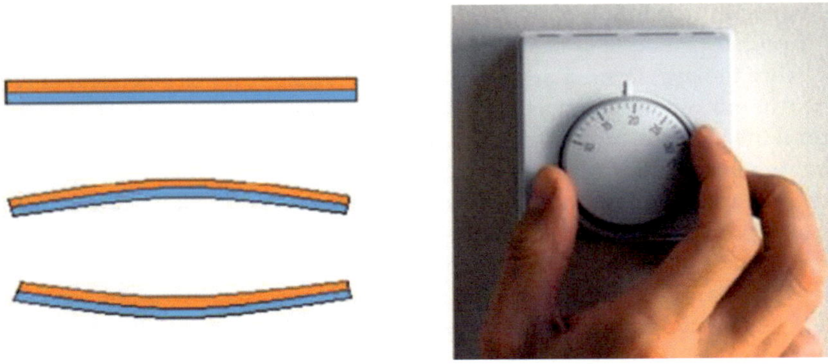

Figure 1.
The principle of bimetal displacement based on temperature (the thermostat), could at one time have been considered "smart" technology. Source: [6].

2.1 Smart materials: details

Polymer chemistry advancements have become part of our everyday lives [1]. The addition of long-chain water soluble polymer lubricant strips have helped in the act and art of shaving by allowing a renewed application of lubrication in every stroke. The apparent structure of the strip is perceived as a magic piece of plastic to most users as they just wet the surface and it continually provides lubrication over hundreds or thousands of shaving strokes (**Figure 2**). The material in the form of polymer molecules forms chains that transfer from the strip to the skin, which provide a smooth surface coating for the razor blades to float over while the cutting surface removes the unwanted hairs, therefore reducing pain during shaving.

Advancements in polymer research have also provided the ability to create specific molecular features or attributes that can be manufactured to provide various states or forms. One type of polymer that can become "smart" utilizes the property of thermochromatic transition states (**Figure 3**). This technology is seen in the previous example of the children's thermometer film but can also be used to create color changes in items dating back to the mood ring of the 1970s. In today's world we can utilize these properties to allow polymer particles to become transparent

Figure 2.
Lubricant strips on razors demonstrate polymer chemistry advancements that have become part of our everyday lives. Source: [11].

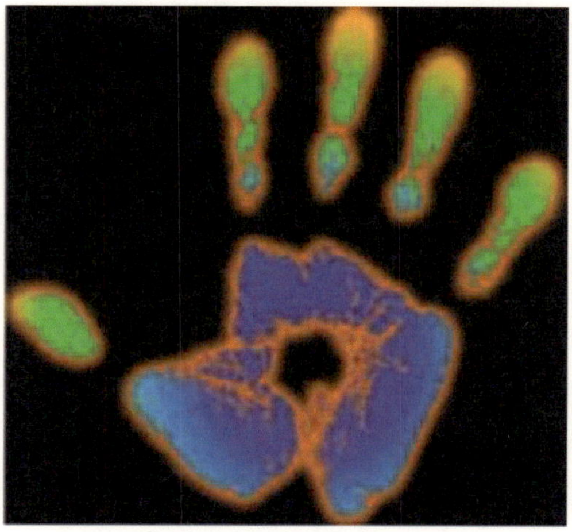

Figure 3.
One type of polymer that can become "smart" utilizes the property of thermochromatic transition states. Source: [1].

based on temperature rise or temperature fall. This can be used to create multilayered signage including building materials that can change color or become invisible.

Another version of smart materials based on temperature is a hybrid material application that utilizes conductive electrical current flow to visually indicate a quantitative energy meter directly on the side of a battery (**Figure 4**).

Thermochromic inks can be used along with a resistive film coating. The display will change based on a battery's voltage. The energy must be sufficient to create a proportionate current flow through the film in order to generate enough heat to display 100% full. The lower the voltage—the lower the bar graph as seen by the observer. Smart particles provide a power meter with no computer involved. Piezoelectric elements can be formed as films, wafers or crystalline particles that can be used as sensors and that can also become actuators (**Figure 5**). The physical properties of the surface can be energized to move based on the voltage level and polarity of the stimulus provided differentially across the surface of the crystal. This property can be used to make a movement, a beeper or a speaker. Another interesting property of the material is that the inverse operation is possible as well. This means that the same device that is a speaker can also operate as a microphone or a displacement vibration sensor. It can also generate small amounts of energy during vibration to yield energy harvesting.

This brings an interesting property of a device called a Peltier device [7] or thermoelectric cooler. The primary function of the special silicon junction is that you can provide a voltage potential that creates a current flow, and the material attributes create a cooling effect on one side of the device while creating a heating effect on the other side (**Figure 6**). An inverse attribute is that the device can also generate a small amount of energy if you provide a temperature differential with sufficient heat and delta temperatures. This feature can be used as a form of energy

Figure 4.
Smart particles provide a power meter directly on the side of a battery. Source: [11].

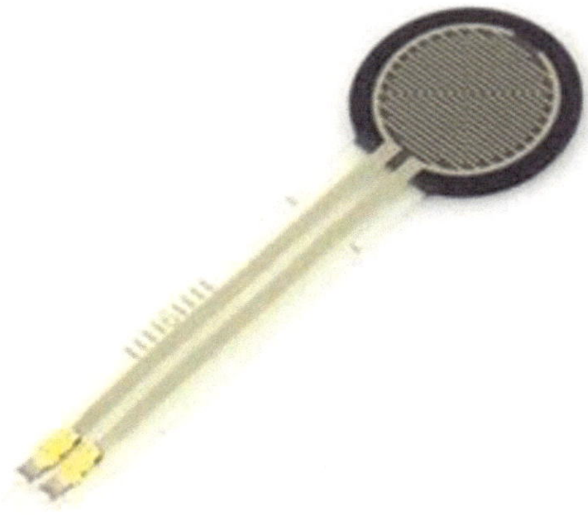

Figure 5.
Piezoelectric elements can be formed as films, wafers or crystalline particles that can be used as sensors and that can also become actuators. Source: [1].

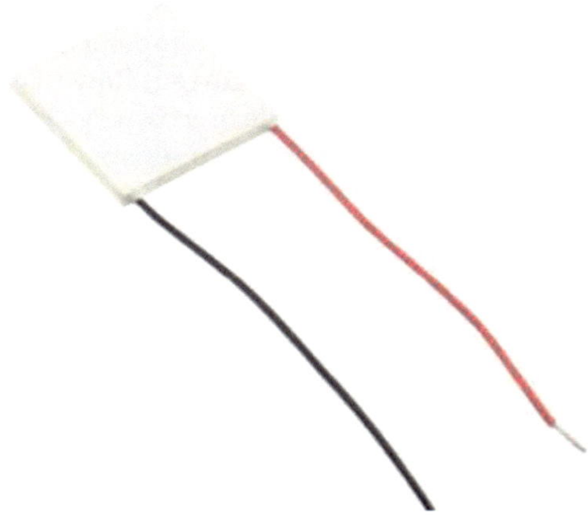

Figure 6.
The special silicon junction of the Peltier device can provide a voltage potential that provides a current flow that creates a cooling effect on one side of the device while creating a heating effect on the other side. Source: [1].

harvesting in the presence of differential thermal surfaces. Organic LEDs are providing the ability to provide an electrical stimulus across a junction of material that can create a transmission of light at a specifically programmed wavelength (or color) (**Figure 7**). An interesting part of this material advancement is that the particles can be manipulated in such a way to create a nonplanar or flexible surface. This will likely evolve into silk-screened TV screens and spray-on digital signage as active billboards.

Electronic ink film is an optical memory technology that can be formed as a grid of static programmable electrochromic (SPEC) particles that are less than 100 microns each that can be used to create sensors, flexible signage, labeling and screen displays. One of the unique properties of the SPEC particles is that they can become reflective or light absorbing and remain in that state semi-permanently without the need for power. Electrical signals in the form of pulsed waveforms are used to excite the particles into discretely programmed states of black, white or a variety of colors (**Figure 8**). The display film is flexible and can be reprogrammed down to the pixel

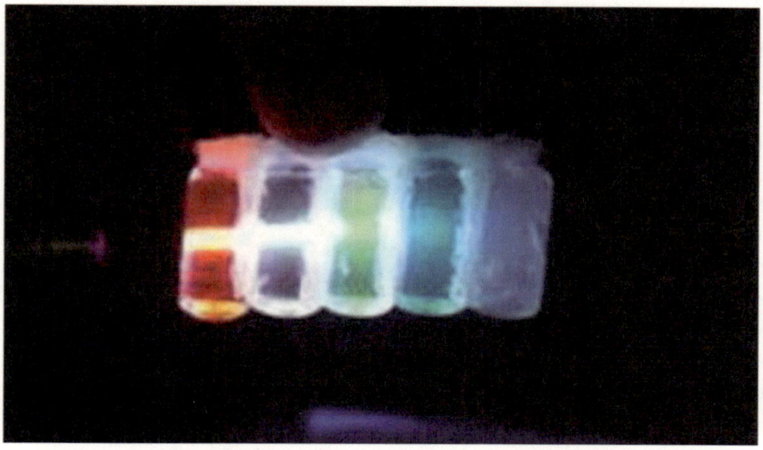

Figure 7.
Organic LEDs can provide an electrical stimulus across a junction of material that can create a transmission of light at a specifically programmed wavelength (or color). Source: [1].

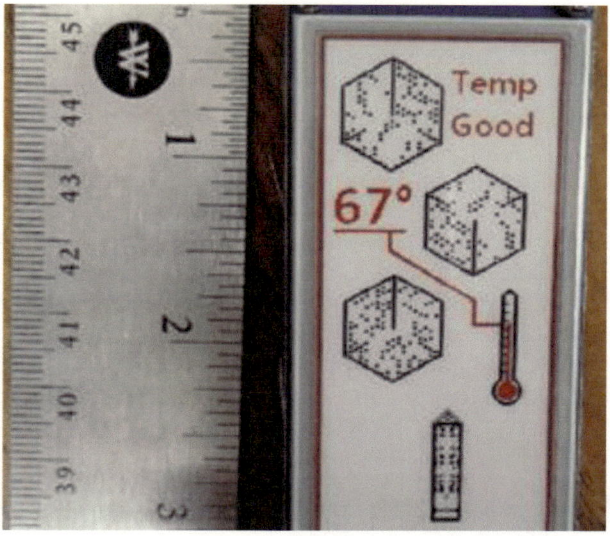

Figure 8.
Optical memory with static programmable electrochromic (SPEC) particles formed as a grid of dots that are less than 100 microns each that can be used to create flexible sensors and nonvolatile labels. Source: [11].

level. There are also segment and block modes for the particles to allow large areas to change colors or thermal properties such as reflective or black body radiating thermal collectors.

Thermochromic polymer particles can be used to coat a surface such as asphalt shingles to allow color changes to the surface-applied particles based on temperature.

Figure 9 shows an example of 90°F trigger programmed particles that reflect light in the summer and turn black in the cooler months in order to absorb heat in the winter. The purpose of the coating is to allow modification to existing production methods while providing a variety of color mixtures and patterns. It is also possible to create other building material solutions using this methodology including inside walls, ceilings and counter surfaces. In some cases the material itself could be injection molded or extruded to provide a functional composite material solution.

The hybridization of these and other smart materials is making its way into other materials such as paints and coatings. One example of this is the recently discovered state of carbon known as graphene, which allows structural enhancements that can be mixed with a carrier component and grown onto a substrate. The end result can form a lattice structure that creates extremely strong bonds structurally as a hexagonal grid. One of the important properties of this material as a lattice is the ability to conduct electricity. This means that the advancements of sensors and computing devices are abundant.

One of the most exciting areas of research into smart materials is in the field of carbon forms such as Fullerene C-60 [8] (Buckyballs), carbon nanotubes (CNT) and graphene.

Figure 10 shows a suspension of CNT that can be used as a coating, paint or adhesive when mixed in production with an appropriate carrier material. Carrier materials such as silicone rubber, epoxies and acrylic paints can provide lattice structures that enhance elasticity, impact resistance and resonant tolerance such as damping properties.

The electrical conductive properties of the carbon additives can create a hybrid material that can be used to create sensors and actuators to tune or de-tune resonant vibrations.

The hybrid composites can also be used to change durometer and response to shock through electrical and/or thermal excitation. Thermal heating modes can create "distributed warming" by conducting an electric current through the hybrid composite to create solutions such as avoiding brittle carrier material characteristics in the cold.

Plastic and epoxy carriers can be formed and molded into strategic shapes based on application or to exploit physical vibration characteristics based on strategically generated waveforms for material displacement. Applications include sound damping and active noise canceling.

Flexible silicone adhesives, gaskets, washers and o-rings can be easily formed to provide programmable response to vibration, pressure, temperature or other electromagnetic properties. Externally powered warming caulk can be created that can offset humidity-based fogging on glass surfaces or defrost cold surfaces.

Figure 9.
Thermochromic polymer particles can be used to coat a surface such as asphalt shingles to allow color changes to the surface-applied particles based on temperature. Source: [11].

Figure 10.
A suspension of CNT that can be used as a coating, paint or adhesive when mixed in production with an appropriate carrier material. Source: [11].

Carbon nanotubes form a contiguous lattice structure during the curing process of the carrier that provides an evenly distributed electrical network throughout the hybrid composite material. One of the inherent properties of CNT lattice structures is the ability to absorb electrostatic discharges (ESD), which can destroy electronic devices. This makes hybrid composite CNT smart materials attractive as electrical safety improvement solutions. Applications such as CNT coatings and potting compound adhesives can become a solution for thermal dissipation as well as energy discharge surface enhancements. Carbon nanotubes and graphene can be structured together in a lattice formation in a complementary manner that can form electrical sensory networks including the formation of a function field effect transistor (**Figure 11**). With this ability to become a smart material hybrid functioning as a transistor we enter a new frontier of semiconductor research that takes smart materials beyond the perception of spectacular to a true functioning computer circuit within the materials of tomorrow.

2.2 Smart adhesives: feature potential

Smart adhesives will someday allow the sensory monitoring of their attributes in real-time and additionally provide a means for tuning (or de-tuning) the properties toward an ideal operating state [1].

Thermal properties are critical to many applications and temperature can affect many of the other attributes listed above. Therefore, much research and development focus is placed on the ability to sense or observe temperature as an absolute reading or a differential reading between surfaces. Additionally, the controlled response mode is desirable to warm or cool mating surfaces (or the adhesive core itself).

Resonance is another important attribute topic for the reinforcement or prolongation of vibrations by reflection from a surface or by the synchronous vibration of a neighboring object [9]. The ability to absorb vibrations can extend life and avoid failure modes to adjoining components. This functionality can be referred to as damping.

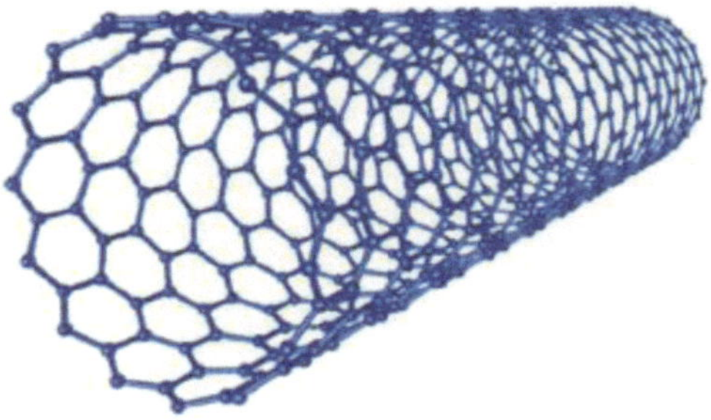

Figure 11.
Carbon nanotubes and graphene can be structured together in a lattice formation in a complementary manner that can form electrical sensory networks. Source: [1].

Damping adhesives have the ability to absorb sound and/or vibrations in such as ways as to isolate the vibrations from one mating surface and attenuate the force seen on the other mating surface. In some applications it may also be desired to absorb or move the vibrational energy throughout the adhesive material itself in order to extend the life of the adhesive structure and bonding properties of the system.

Vibration can be an attribute that is quantified as resonance where energy over a period of time reaches a repeating cyclical pattern that can become destructive (or in some cases can be desirable). The resonant frequency of an object is commonly measured by the number of vibrations per second and can be represented as Hertz. The ability to vary or alter energy cycling is based on measurable components of amplitude, fundamental frequency, frequency harmonics, thermal loss and transfer of vibration known as propagation.

The force experienced by a surface can be transferred into the adhesive's mating surface through layers or composite core material(s) and ultimately propagate onto the adjoining alternate surface. Motions within the adhesive core materials (and layers of the adhesive) will either transfer energy, absorb energy through direct motion or convert the conducted energy into another energy form. The natural form of energy for conversion by default is generally heat. The small amount of heat energy experienced during vibration can create a reduction (or attenuation) of vibrational energy between mating surfaces. Therefore, the addition or removal of heat may be a helpful method of altering the propagated (or conducted) vibrations between surfaces.

The concept of resonance is important beyond the simple idea of vibrating surfaces. Every object or grouping of objects has a specific fundamental frequency of tuned resonance as well as affected responses at harmonic frequencies. Harmonic frequencies are similar to notes on a piano or guitar string. The fundamental frequency is the specific vibration in Hertz that may cause damage to your mating surface components or the adhesive composites themselves. Much like a chord in music having a 1, 3, 5 harmonizing tone, the vibrations will yield other ripples of energy within the adhesive at other frequencies beyond the fundamental tone (or vibration). You can think of this as ripples seen in water as you drop a rock into a lake. The ripple patterns created are more complex than a simple wave of energy.

If the adhesive can absorb certain movements over time it is possible for the peak amplitude of the resonant frequency to be reduced and ultimately smooth out the overall vibrational energy that would cause failure modes to the mating bonded system. The ability to monitor and sense this vibrational energy and allow reduction of amplitude can be a powerful method of extending life to components and bonding attributes. Spreading energy over time and converting vibrational energy to other states are some of the techniques in creating a damping adhesive.

2.3 Damping adhesives

One of the goals of a damping adhesive is the ability to compress and expand layers of material while bonding adjoining surfaces [10]. One method of achieving this goal is to layer or embed components into a carrier where vibrational energy received by the substrate causes particles to compress from a natural expanded state to a compressed state. The composite layer components subsequently return to their expanded state, thereby releasing an opposing energy force onto the substrate which is slightly less than the initial applied energy received by the substrate. The remaining energy loss is in the form of heat through frictional vibrations.

More advanced modes of a damping adhesive system can be realized. A controlled energy field in response to a sensed stimuli can be applied to minimize vibrations. The layered damping system could include a sensory layer which measures an amplitude and frequency spectrum of the disruption. The sensor determines the amplitude and frequency spectrum of the disruption received by the substrate; and the applied force field is dependent on the amplitude and frequency spectrum of the disruption. The overall goal of this composite design is to Tune or Detune resonant frequencies of the overall adhesive bond.

It is also possible to create a vibrational response through thermal or electrical patterns that can provide an opposing force to the energy disruptions by vibrating an opposite (or inverse) waveform to noise-cancel vibrational energy. This is much like the performance and purpose of a sound-canceling headphone effect across two surfaces of adhesive bond layers or substrates.

2.4 Functional overview

Each adhesive serves a particular purpose, whether it be the bonding of metal panels in a Navy ship where the selected adhesive is chosen based on its ability to firmly bind substrates together with a low likelihood of failure in less than ideal condition, or to affix parts of an electronic device together where the selected adhesive is chosen based on its ability to allow the materials to be bonded together but still flex to try to minimize breakage of the component parts. For example, everything composed of matter is in a constant state of fluctuation and has a resonating frequency. If a given object is vibrated at its resonant frequency, or even a harmonic of that frequency, the object may be destroyed as a result of power amplification (i.e., power that is built-up over time at a precise frequency (the object's resonant frequency). A very small amount of applied energy (i.e., forcing frequency) can thus cause destruction of a large object. Frictional forces (e.g., dampers) may be applied to an object to slow the motion of the resonating frequency and attenuate the amplification. Because adhesive is found in nearly every product, it may be an ideal substance to act as a damper. Indeed, the elastic nature of certain types of adhesives may, by themselves, act as a damper. However, the damping effects of the adhesive may be limited by the characteristics of the adhesive itself. Accordingly, an adhesive that has an increased ability to act as a damper in addition to an adherent is desirable.

A substrate may experience a physical shock, which may have detrimental effects. While an adhesive having flexible properties may be effective to reduce some of the effects of the shock, the adhesive may have limited flexible behavior and over time may begin to fail, becoming more brittle and therefore less able to deflect or dampen the shock such that the substrate remains undamaged. Accordingly, an adhesive having an increased ability to deflect or dampen a physical shock received by a substrate in addition to acting as an adherent could be helpful.

For example, an adhesive with a vibrational damping particles, for reducing the effect of energy transfer to or through one or more of the substrates to which the adhesive is applied. Damping particles could include particles of different sizes and shapes, and may include nano-particles, micro-particles and macro-particles. In some cases the damping adhesive material may consist of a combination of multiple types of coordinating adhesives which may be strategically selected and provided as a layered medium (e.g., silicone, polyurethane, epoxy, latex, etc.).

Damping particles could potentially contain spokes and may be formed from a material exhibiting superior flexibility and elasticity, such as thermoplastic polyurethanes (e.g., TPU 92A-1). Thermoplastic urethanes may exhibit durable elasticity, high resistance to dynamic loading, high abrasive resistance, quick response and good temperature range. Alternately they may be formed of a material that exhibits greater stiffness than the spokes and optionally interact with elements that may be formed of a material that exhibits less stiffness than the spokes. It may also be desirable to include particles that are magnetic such as iron/steel filings or ferrofluid. The goal of the varying durometer and ingredient layers is to provide a mechanism for monitoring internal forces and optionally apply dynamically controlled responses.

Some of the damping particle materials may be used to form bonds that have a tendency to remain in a naturally expanded state. When a change in the environment of the bonded substrates occurs, e.g., due to an applied energy on the substrate(s) from any direction, the applied energy causes the bonded structure to temporarily flex or compress. As a result of the compression of the structures, some of the stress to the substrates is diffused from the substrate and transferred to the structures. The structures may eventually return to their natural expanded state, and in doing so, return an opposing applied energy to the substrate(s). The opposing applied energy returned to the substrate may be less than the original applied energy. However, due to the structures' ability to diffuse some of the applied energy from the substrate, the substrate may remain relatively undisturbed.

The amount of compression experienced by the structures may be directly related to the strength of the applied energy upon the substrate. The applied energy could be the result of any type of disturbance to the environment, including but not limited to sound waves, electromagnetic waves, seismic waves, changes in temperature and/or pressure, physical shocks, etc. In the instance of a physical shock, the applied energy received by the substrate, and thus the structures, may be substantially greater than the applied energy received as a result of sound waves. Therefore, the bonded structures may experience a greater degree of compression in order to diffuse the energy from a physical shock than they would to diffuse the energy from sound waves.

For example, a smart adhesive may be used for applying ceramic shingles to the roof of a building. The adhesive may be applied to the underside of the shingles and the shingles applied to the roof. During a thunderstorm, the roof may experience hail, which exerts an applied energy on the shingles which, in some instances, may be sufficient that the ceramic shingles would traditionally crack. However, due to the enhanced adhesive having damping adhesive, when the shingles receive an applied energy from the hail, energy is at least partially transferred to the bonding structures such that the structures flex or compress. In this case, because the energy may be greater, the structures may experience a greater amount of compression. The structures then return to their natural state due to the elasticity of the structures, which returns an opposing applied energy to the shingle. Having diffused some of the original applied energy, the opposing applied energy that is returned to the shingle is less than the original applied energy. Accordingly, due to the transfer of energy to the structure, the shingle is less likely to crack.

In another example, a damping adhesive bonding structure may be applied around the outer perimeter of a window. Sounds waves traveling through the air hit the window (e.g., on an outside surface), and traditionally would travel through the window to the other side (e.g., an inside surface). However, due to the damping adhesive, when the window receives energy from sound waves, the energy is at least partially transferred to the damping bonding structures. In this case, because the energy may be relatively small, the structures may only experience a slight degree of flex or compression. The structures then return to their natural state, thus returning an opposing applied energy to the window. Having diffused some of the original applied energy, the opposing applied energy that is returned to the window is less than the original applied energy. As a result of the diffusion of energy by the structures, the sound waves traveling through the window may be considerably decreased.

Damping particle bonding structures can be partially constructed utilizing embedded damping particles with adhesive carriers that take the form of a spheroidal molecule, geodesic dome or other three dimensional shapes. Many such particles exist in nature, or have previously been developed, for various applications. Fullerenes are one example of a damping particle. Fullerenes are a class of allotropes of carbon which are essentially sheets of graphene which can be rolled into tubes or sphere. One example of a fullerene molecule, C_{60}, comprises 60 carbon atoms arranged as 20 hexagons and 12 pentagons to form a soccer ball (or buckyball) shaped structure. Graphene, for example, may be also be provided as a box-shaped structure (e.g., as a layered structure). Carbon nanotubes can be used also where the graphene molecules have been rolled into a 3 dimensional tube. A suspension of tubes in an adhesive may preferably result in the nanotubes being dispersed in various orientations, e.g., some oriented vertically, some oriented horizontally, and some oriented at various angles. In this way, the nanotubes may be effective to dampen applied energy received by any substrate at any angle.

A dendrimer is another example of a damping particle. Dendrimers are spherical polymeric materials whose properties are usually determined by functional groups appearing on the molecular surface. Dendrimers may be used in the synthesis of monodisperse (i.e., uniform) metallic particles. Poly(amidoamine) dendrimers are often used, and the end result may be a dendrimer-encapsulated particle.

The spheroidal, geodesic dome may function substantially similarly where, in the case of C_{60}, the bonds between carbon atoms may have some degree of flexibility which allows the molecule to flex or compress when energy is applied to the molecule.

It is also possible to exploit opposing forces for damping by means of magnetism. It may be preferable for the damping particles to exhibit magnetic properties. In these cases the extended benefit of electro-magnetism may allow damping adhesive response through modes that are electrically or electromagnetically active. When suspended in an adhesive, the adhesive may take the form of a ferrofluid or a liquid (liquid, gel, etc.) that becomes magnetized in the presence of a magnetic field. Ferrofluidic adhesives allow for passive and/or dynamic response to applied energy to a substrate.

Ferrofluid is a unique material that acts like a magnetic solid and like a liquid. In this case, by incorporating damping bonding structures having magnetic properties into an adhesive, the adhesive may be transformed into a ferrofluid. A ferrofluid is superparamagnetic, allowing the liquid to display magnetic tendencies only in the presence of a magnet. Thus, in order to transform an adhesive from a liquid to a ferrofluid, the damping apparatus must have magnetic properties.

Carbon-based particle structures, such as C_{60}, may be naturally paramagnetic, i.e., behaves like magnets in the presence of a magnetic field. Other structures, such as those manufactured from a polymer, may be coated in, or otherwise incorporate a magnetic material such as iron oxide. It may be desirable for the magnetic material to be coated in a surfactant to keep the magnetic structures from sticking together.

Absent a magnet, the particle structures may function as described above. In other words, without a magnet, the damping apparatus may simply compress as a result of an applied energy, thus diffusing some of the applied energy away from the substrate. In the presence of a magnetic (or electric) field, however, the adhesive may become an even more effective damper.

A magnetic field (e.g., using a magnet or a magnetizable material, such as a small rod of iron alloy wrapped in a current-carrying coil) may be applied evenly at or near the areas of the substrate having the adhesive to influence the orientation of the damping apparatus. This may be most useful in the case of an adhesive suspension comprising tubes. For example, if an applied energy most frequently occurs in a single direction across the substrates, then the magnetic field may be applied such that the tubes are oriented so as to transversely receive the applied energy.

An applied energy field may additionally be received by the substrate in one or more concentrated areas. In this case, a magnetic field may be activated only in the concentrated area(s) receiving the applied energy. Thus, multiple pieces of magnetizable material may be provided at or near the adhesive areas such that multiple magnetic fields may optionally be applied. The magnetic damping particles will thus be drawn to the magnetic field(s). An increased concentration of the flexible damping adhesive substrate may thus provide increased flexibility in the area of the magnetic fields.

The magnetic field may be turned on, or off, but the magnetic field may not vary, for example, over time, or in response to dynamic changes in the substrate environment. This method can be accomplished through high-frequency modulation techniques.

As an applied energy, or force energy, is encountered by the substrate, the sensors analyze the applied energy to ascertain the amplitude and frequency spectrum of the applied energy. A force field may be applied at or near the adhesive having damping apparatus dispersed therein to alter the properties of the damping apparatus (e.g., change in orientation, elasticity, etc. of the damping apparatus) in such a way that the damping apparatus experiences a degree of physical displacement at controlled timing intervals based on the frequency spectrum of the applied energy. As an example, the damping particles may actually experience controlled oscillations (e.g., physical displacement along a particular distance). The controlled response of the damping apparatus may result in a response force that is in a spread spectrum inverse waveform which may geometrically stabilize the substrate to avoid peak resonant frequencies which may damage the substrate(s). In other words, the controlled adjustments and corresponding response of the damping apparatus may result in a decrease of the amplitude of the applied energy by spreading the applied energy out over time.

For example, one or more sensors may be placed at or near a substrate (e.g., a window) for detecting sound waves. During peak hours of the day (e.g., high traffic times) the sensors may register higher decibels of sound waves being transmitted and received by the window; conversely, during the evening hours, the sensors may register lower decibels of sound waves. In response, the amount of current pushed to the magnetizable materials may be greater during the day in order to block out unwanted noise than that required during the night.

In another example, a sensor may detect the natural resonance frequency of the substrate. A second sensor (or the same sensor) may detect the frequency of applied energy upon the substrate. If the second sensor detects that the frequency of the applied energy is the same as the natural resonant frequency of the substrate, a magnetic field may be applied (or removed, as the case may be) at or near the adhesive in order to alter the properties of the damping apparatus such that the apparatus may dampen (or alter) the frequency and amplitude of the applied energy as received by the substrate in order to avoid power amplification and possible destruction of the substrate (or structure attached to the substrate). This ability to tune and detune the damping apparatus in real-time in response to a sensed frequency of a substrate or energy applied to a substrate may allow for supremely customizable products onto which the inventive adhesive is applied. It is important to note that the resonant frequency of the substrate may be altered (or change) by the mass or a surface which may be affixed to the adhesive or substrate.

Applying an electric field (e.g., low voltage pulses) at or near the adhesive substrate to influence the damping structure can allow provide a change in the apparent viscosity or the durometer of the adhesive material. An electric field may be applied at or near the adhesive area. For example, conducting plates may be provided parallel to each other (e.g., at each substrate surface). A voltage may be maintained between the plates by passing current through the plates. The apparent change in the viscosity of the adhesive may be directly dependent on the strength of the applied electric field. Thus, as the strength of the applied electric field is increased and/or decreased, the consistency of the adhesive may transition from that of a liquid to a gel, and vice versa (or to and from a more elastic gel to a less elastic gel).

The change in viscosity or durometer of a material may occur over very small time increments, e.g., milliseconds, making the electrorheological adhesive especially useful in conjunction with sensors. For example, the electrorheological adhesive may be applied to a window. The conducting plates may optionally be opposing sides of the sash, if the sash is constructed of, for example, aluminum. Alternately, conducting plates may be provided parallel to each other on either side of the sash. The window pane(s) may be placed between the plates. One or more sensors may be placed at or near the window pane(s) to measure applied energy to the window pane(s). If the sensor senses an applied energy over a threshold value, the sensor may transmit a signal to cause an electric field to be applied to the plates. In response, the adhesive (via the damping apparatus) may become stiffer in order to reduce the effects of the applied energy.

The damping particles may also take the form of piezoelectric particles. In response to an applied force, the piezo particles become deformed. For example, when the piezo particles are deformed in one direction due to an applied force, a force field (e.g., voltage impulse) is activated (e.g., via a signal from a sensor) to send electric power to the piezo particles to bend in the other direction. In this way, the response of the piezo particles can help to reduce the disruption to the substrate.

Variations of arrangements including combinations of hybrid configurations of smart-materials such as nickel titanium or nitinol can be combined to perform dynamic response by utilizing the properties of shape-memory alloys (SMAs). SMAs and other smart-materials may exhibit properties that allow the molecular alignment to change in physical state (i.e., relative molecular position) based on variations in energy levels experienced by the SMA material. For example, in the form of a wire strand, nitinol varies in length based on the temperature of the SMA itself. In the case of conductive smart-materials such as nitinol, an electric current can be induced into the smart-material/SMA in order to alter the temperature of the SMA-causing the length of the wire to vary based on changes in the wattage dissipated across the SMA wire within the adhesive structure.

Variations of state, position and structure within SMAs and smart-materials can be exploited to embed SMA wires, pellets or thin-film strips within the adhesive structure. By varying the SMA's density, position and relative placement between companion particles the effective resonant mode of the adhesive can be changed based on external stimulus as a controlled response to achieve anti-resonant damping.

SMA pellets can be dispersed within a mixture of particles to create a layer which may have various properties (e.g., fluid, gel, plasma, etc.) that can be altered in dimension based on externally induced waveforms. A layer may be strategically positioned between two other layers of, for example, pressure sensitive adhesive, which may be equipped with conductive strips or foil plates. Direct contact of the external response waveforms or force fields can be conductive as direct current (DC) or low-frequency. Indirect induced waveforms can also be used to capacitively couple electrical energy through the SMA pellets using strategically selected high-frequency alternating current (AC) waveforms which do not require direct electrical contact. One example of an indirect method of induced waveforms is the use of ultra-high-frequencies that are electromagnetically propagated in the form of short-burst pulse streams that are strategically shaped in amplitude, wave shape and frequency to achieve selective resonance of specific particle layers or positions of particle regions along or within a substrate layer. The short-burst pulse streams are managed over time in frequency bursts where the frequencies are generated in alignment with wavelengths and fractional wavelength timings (such as quarter-wave, half-wave and full-wave periods). These methods can be used to achieve selective areas of resonant tuning (and detuning) in order to vary the durometer of the adhesive layer(s) in real time. In other words, a nearby field of energy can be utilized to vary the anti-resonant damping mode of the adhesive without any direct contact to the adhesive itself.

Damping adhesives may be extruded, for example, and may have a muscle wire (or SMA wire) located within the extrusion. Electric current may be conducted through the wire via contact leads.

Smart-materials can be somewhat slow to respond due to the thermal mass or other physical properties that can slow response time. This means that there is a limit to the response time (or frequency) of the molecular changes in the smart-material (or SMA) itself. One method of obtaining increased performance is to utilize a harmonic frequency byproduct (based on the changes in the physical properties of the SMA material) to assist in the anti-resonant damping process. For example, a change in SMA structure may be possible in 100's of milliseconds occurring in a repetitive pattern at a fundamental frequency (or rate of change) altering the SMA structural alignment. A resonant byproduct of this movement-pattern can be realized by strategically using the 3rd (or 5th, etc.) harmonic with notable energy that can be used to assist in the anti-resonant tuning and detuning of the adhesive structure for damping. By utilizing a higher frequency harmonic as the controlled response damping, you can achieve this result by providing a much lower rate of change to the molecular smart-material/SMA and achieve higher frequency movements in the substrate structure for damping. The net result allows slow movements within the adhesive structure to provide damping to higher frequency vibrations which in turn enhance damping performance of the adhesive.

2.5 Smart conformal coatings

Adhesives, which may be configured as curable coatings which lose tackiness upon curing, have been used as coatings to protect and isolate surfaces of objects. Moreover, adhesives have traditionally not been equipped with means for providing real-time information about the objects to which the adhesives are adhered. Once the adhesive is applied, it is forgotten about, and further information regarding the objects may be determined using other means.

A smart adhesive having damping properties could be used as a conformal protection coating and sensory material health monitoring surface. A damping system for reducing the effects on a substrate caused by a disruption in the substrate environment could include an adhesive having a damping particles dispersed therein. The particles could be configured to provide a controlled response to an applied force field. The conformal coating system includes a sensor which measures an amplitude and frequency spectrum of the disruption. The sensor determines the amplitude and frequency spectrum of the disruption received by the substrate; and the applied force field is dependent on the amplitude and frequency spectrum of the disruption.

2.5.1 Conformal coating example overview

Adhesive applications may sometimes take the form of a coating, such as a paint or lacquer, although coatings may be any covering that is applied to the surface of an object. Coatings may be used on many different objects to provide benefits thereto. A damping and/or information-transmitting coating may be particularly useful at locations that are relatively difficult for a person to access. One example of such a location, or rather a plurality of locations, is at a pipeline.

A smart pipe coating utilized to adhere sections of pipe together, to locate possible weak spots in the pipe, locate electrically-exposed areas of the pipe, locate ruptures in the pipe and/or provide enhanced communication capabilities along the length of the pipe. One or more pipe sections adhered together at a point of abutment. Generally, the pipe sections may include a machined end designed to fit inside the other pipe section.

Occasionally, the pipe sections are coated with a substance designed to prevent or slow corrosion of the pipes. The coating is designed to prevent moisture in the ground from coming into direct contact with the pipes. Usually, the coating process is completed before the pipe sections are delivered to the construction site. However, the coating is not applied to the ends of the pipes in order that the coating does not interfere with the welding of the pipe sections. Once the pipe sections are welded together, a coating crew coats the remaining portion of the pipe in the coating, and the pipe is lowered into a prepared trench.

The coatings currently available are one-dimensional in that they are designed only to protect the pipes, but do not serve additional functions such as preventing unwanted vibrations, helping to locate potential weak spots in the pipes, identifying locations of ruptures, or increasing communications along the length of the pipe.

A smart adhesive coating could be used as a coating for a pipeline. The coating composition may include bonding particle structures or particles for reducing the effect of energy transfer to or through a pipe. The ability of the bonding particle structures to resist or reduce energy transfer through the pipe as a result of forces acting upon the pipe from both the inside and the outside of the pipe may help to extend the life of the pipe. Programmable sections of the pipeline may be selected to allow variable conductivity from an excitation source onto a desired pipe section in order to provide dynamic cathodic protection (DCP).

The bonding smart particles may be magnetic, and/or electrically or electromagnetically active. The coating may therefore be configured to provide a dynamic response to a force happening upon the pipe. Sensors may be utilized to measure the amplitude, frequency and wave shape of the waveform of the force happening upon the pipe. The sensors may additionally measure environmental attributes, such as soil content (e.g., alkalinity, salinity, temperature, pressure, etc.). Using magnets, electromagnets, electrical pulses, etc., the particles may be persuaded into a position, pattern or oscillation. The persuasion or oscillation of the particles may attenuate the waveform, or present an alternative waveform. The alternative waveform may be inverse to the waveform of the force happening upon the pipe. DCP techniques may be incorporated to prevent corrosion of pipe sections by forming a closed-loop of sensory and dynamic controlled response to provide a superior form of cathodic protection to existing passive (or fixed mode) techniques.

2.6 Smart adhesive research focus

2.6.1 Carrier and subcarrier hybrids

Direct specific research areas have been focused on smart damping adhesive (SDA) applications as gaskets, o-rings seals, caulks and conformal coatings [10]. One of the important concepts in creating a practical smart adhesive solution is the concept of layering films, sprays or laminate surfaces to build-up a composite solution as a smart adhesive. **Figure 12** shows an SDA pad that can thermally warm two surfaces and monitor strain between the two surfaces.

Smart adhesive layer attributes include:

- Bonding properties
- Optical filtering
- Electrical conduction
- Electrical insulation
- Vibrational damping
- Thermal dissipation
- Thermal insulation
- Durometer variations
- Viscosity variations
- Shape memory
- Transducer functions

Research with layers in silicone as a planar sheet, film and spray have shown test data that can provide electrical properties as follows:

Figure 12.
Smart damping adhesives can be formed as pads, gaskets, seals, o-rings or other three dimensional shapes. Source: [11].

Note: utilizing various mixtures of room temperature vulcanizing (RTV) silicone with carbon nanotubes the following conductivity was achieved:

Electrical layer properties (Ohms per square):

- 1 Meg Ohm for electrostatic discharge layer
- 10–20 K Ohm for pressure/strain-gauge displacement layer
- 10 Meg Ohm for optical transparency layer
- 50–2000 Ohm for thermal excitation layer

Another important topic is the ratio mixtures for a carrier adhesive component such as Silicone (or Epoxy) is the percentage addition of sub-carrier materials such as carbon nanotube or graphene. Note: The mixing methods are currently experimental and meant to establish proof of concept for each layer.

Other carrier components include piezo particle layers for vibrational sensing, controlled frequency response and energy harvesting. Future research will incorporate embedding subcarriers of piezo particles into silicone, epoxy and plastics such as mylar.

The resulting research test gaskets, o-rings and sprays exhibit consistent results and should have a promising future potential in the world of smart adhesives.

Industries have many potential strengths to embrace by utilizing the future benefits of smart adhesives in the future.

3. Conclusion

3.1 Smart adhesives summary

Smart Adhesives have the potential to grow into a significant industry segment of Adhesives, Coatings, bonding agents, layers and hybrid chemistry solutions that do not yet exist today. In the future it is possible that we will achieve layered adhesive solutions that can sense, monitor, communicate, and potentially self-heal. There are many uses of thermal, electrical and physical displacement that will become features in adhesives that will extend the life and functionality of the mating objects as well as the adhesive itself.

The advancements in microencapsulation, film layering, polymer research and hybrid material matrix design is moving exponentially and will likely be a part of your future.

Acknowledgements

The research into smart adhesives is a long and continuous journey. This journey would not be possible without the contributions and support of many people. I would like to acknowledge a few of the key people involved in this effort.

Fielding Staton, CEO, Co-inventor, WINDGO, Inc.; Tony Ewen, Lead Physicist, WINDGO, Inc.; Justin Eikel, Mechanical Engineer, WINDGO, Inc.; Mason Pramod, Chemical Engineer, WINDGO, Inc.; William Whitacre, Software Engineer, WINDGO, Inc.; Jim Kyd, Research Specialist, WINDGO, Inc.

Appreciation and thanks

I would like to thank the following special people. Your contributions, efforts, support and tolerance are highly appreciated and this would not have happened without you.

Meg Ladd, Fielding Staton, Anna Quinn, Justin Poplin, Tony Ewen, Justin Eikel, Mason Pramod, William Whitacre, Jim Kyd, Dr. Susan Burkhart, Lee Strumpf, Darla Staton, Dr. John Miles, Dr. Bill Buttlar, Zane Browning, Lili Vianello, John Shrum Dr. Ron Frederick, Dr. David Stalling.

Author details

David Strumpf
Vice President of R&D, WINDGO, Inc., Columbia, MO USA

*Address all correspondence to: d.strumpf@windgo.com

References

[1] Strumpf D, Finding a Path to Smart Paints, Adhesives, Films and Alloys using Smart Materials [Internet]. 2018. Available from: https://www.pcimag.com/articles/104421-finding-a-path-to-smart-paints-adhesives-films-and-alloys-using-smart-materials [Accessed: 2019-03-06]

[2] Nickel Titanium—Nitinol [Internet]. 2018. Available from: https://en.wikipedia.org/wiki/Nickel_titanium [Accessed: 2019-03-06]

[3] Smart Material Definition [Internet]. 2018. Available from: https://en.wikipedia.org/wiki/Smart_material [Accessed: 2019-03-06]

[4] Designing a Nanotube Using Naturally Occurring Protein Building Blocks [Internet]. 2006. Available from: https://www.ncbi.nlm.nih.gov/pmc/articles/PMC1447657/ [Accessed: 2019-03-06]

[5] Galileo Thermometer, History [Internet]. 2018. Available from: https://en.wikipedia.org/wiki/Galileo_thermometer [Accessed: 2019-03-06]

[6] Bimetal strip figure [Internet]. 2018. Available from: https://www.introduction-to-physics.com/bimetallic-strip.html [Accessed: 2019-03-06]

[7] Peltier Effect, History [Internet]. 2018. Available from: https://en.wikipedia.org/wiki/Thermoelectric_effect [Accessed: 2019-03-06]

[8] Carbon C60—Buckyball, History [Internet]. 2018. Available from: https://en.wikipedia.org/wiki/Buckminsterfullerene [Accessed: 2019-03-06]

[9] English Oxford Living Dictionaries, Resonance, Physics Definition [Internet]. 2018. Available from: https://en.oxforddictionaries.com/definition/resonance [Accessed: 2019-03-06]

[10] Staton F, Strumpf D. Smart Damping Adhesive II. US patent no. 10,088,011; 2018

[11] Windgo, Inc., www.windgo.com. Copyright 2019. Windgo, Inc

Chapter 3

Epoxy Adhesives

Chunfu Chen, Bin Li, Masao Kanari and Daoqiang Lu

Abstract

Epoxy adhesives are primarily composed of epoxy resin and curing agent. Epoxy adhesives are supplied in both one-component package and two-component package depending on curing agent used and curing method applied. Two-component epoxy adhesives are prepared by packing epoxy composition and curing agent composition separately. They cure soon after mixing the two components together. Almost all room temperature cure epoxy adhesives are supplied in two-component package. One-component epoxy adhesives are prepared and supplied by mixing all formulated components in advance including epoxy resin and curing agent. One-component epoxy adhesives usually need cure at elevated temperature and store at low temperature in a refrigerator or even freezer. Epoxy adhesives have been widely used as typical reactive adhesives for various applications ranging from general industry, construction, electronics assembly, automobile production to aerospace market. Typical room temperature cure epoxy adhesives, thermal cure epoxy adhesives and UV cure epoxy adhesives are introduced in detail.

Keywords: epoxy adhesive, one-component, two-component, room temperature cure, thermal cure, UV cure, latent curing agent

1. Introduction

Epoxy adhesive was first invented in 1936 by Dr. Pierre Castan for dental application via curing bisphenol A epoxy resin with phthalic anhydride. Commercial supply of epoxy adhesives was started in late 1940s in Europe and USA. Various epoxy adhesives have been developed and commercialized since then and widely used as typical reactive adhesives for various structural bonding applications ranging from general industry, construction, electronics assembly, automobile production to aerospace market [1–11]. Their typical application area and examples as well as their supply type, curing method are summarized in **Table 1**. Major global suppliers for epoxy adhesives are Henkel AG & Co. KGaA, H.B. Fuller Company, 3M, Huntsman Corporation, Sika Corporation, Arkema Corporation, Cemendine Co., Ltd., Three-Bond Co., ltd., Huitian Adhesive, etc.

Epoxy adhesives show good adhesion on various substrates and are suitable to bond metals, glass, concrete, ceramics, wood and many plastics. Curing shrinkage is very low. Cured epoxy resin possesses strong and rigid cross-linked chemical structure suitable for structural bonding applications. By combination of various epoxy resins and different curing agents, a number of epoxy adhesives have been commercialized for different applications. On the other hand, room temperature and thermal cure epoxy adhesives need relatively long cure time. Most cured epoxy adhesives are very rigid and are not suitable for bonding flexible substrates.

Application area	Application examples	Package type	Cure method
Industrial	Structural bonding	One component; Two components	R.T. cure Thermal cure
Construction	Concrete repairing Anchor bolt fixture	Two components	R.T. cure
Automotive	Structural bonding Hemming adhesion	One component; Two components	Thermal cure
Aerospace	Metal, honeycomb & composite bonding, repairing	One component; Two components	Thermal cure
Electronics	Electrically conductive Display assembly Image sensor assembly Underfills Medical bonding	One component; Two components	Thermal cure UV cure R.T. cure
Others	Sports tools Consumer applications	Two components	R.T. cure

Table 1.
Typical applications of epoxy adhesives.

In selection and use of epoxy adhesives, cautions need to be paid on their pot life, cure condition, cure method, physical properties of un-cure and cured resin as well as adhesion performance.

2. Epoxy adhesive chemistry

Epoxy adhesives are primarily composed of epoxy resin and curing agent. Filler, toughener, plasticizer and other additives such as silane coupling agent, deformer and colorant, etc., can be formulated as needed. Common compositions and their main role of epoxy adhesives are illustrated in **Table 2**.

Epoxy resins are mainly synthesized from reaction of active hydrogen in phenols, alcohols, amines and acids with epichorohydrin, abbreviated normally as ECH at certain well controlled conditions. Epoxy resin can be also prepared by oxidation of olefin with peroxide as in the case of preparation of cycloaliphatic epoxy resins. Main commercial epoxy resins, their preparation and key features are shown in **Table 3**. Bisphenol A diglycidyl ether, often called as bisphenol A type epoxy resin, is the first commercialized and still most widely used epoxy resin. Synthesis of DGEBA is illustrated in **Figure 1** [12]. In volume base, it is estimated that over 75% of epoxy resin used in industry is this type. **Figure 2** illustrates chemical structure and key features of various functional groups for bisphenol A diglycidyl ether [13], the most common epoxy resin used in epoxy adhesives.

Epoxide group is chemically very active. Epoxy resin can react with active hydrogen almost equivalently via polyaddition mechanism with polyamines, mercaptan compounds, phenols and anhydrates to become cross-linked strong thermoset polymers. Epoxy resin can also polymerize homogeneously via anionic polymerization mechanism by initiating of Lewis bases such as tertiary amines or imidazole compounds. It can also polymerize via cationic polymerization mechanism via initiating of Lewis acid such as Boron trifluoride amine complex or strong acid such as onium salts, iodonium salts. **Table 4** lists typical curing agent, initiator used in epoxy adhesives. By combination of suitable epoxy resin with curing agent, epoxy adhesive is designed for various substrate bonding in different applications. It is supplied in both two-component and

Constituent	Ingredient	Main role
Primary	Epoxy resin, reactive diluent	Adhesive base
	Curing agent/catalyst, accelerator	Curability
Modifying	Filler	Property modification
	Toughener	Toughening
	Plasticizer	Flexibility
Additive	Coupling agent	Adhesion
	Colorant	Color

Table 2.
Epoxy adhesive compositions.

Epoxy resin type	Preparation from	Key features
Glycidyl ether of		
Bisphenol A Bisphenol F Novalac	Bisphenol A and ECH Bisphenol F and ECH Novalac with ECH	Standard epoxy resin Low viscosity Multi-functional
Glycidyl ester	Carboxylic acids and ECH	Mainly for anhydrate cure
Glycidyl amine	Amines and ECH	Multi-functional
Cycloaliphatic	Oxidation of olefin by peroxide	Cationic cure

Table 3.
Commercial epoxy resins.

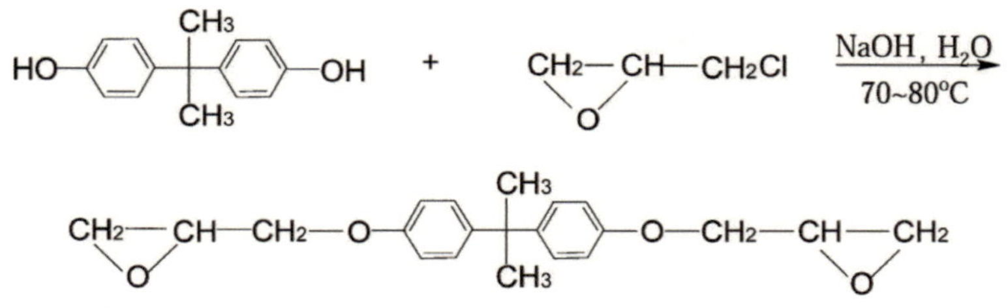

Figure 1.
Synthesis of DGEBA (diglycidyl ether of bisphenol A).

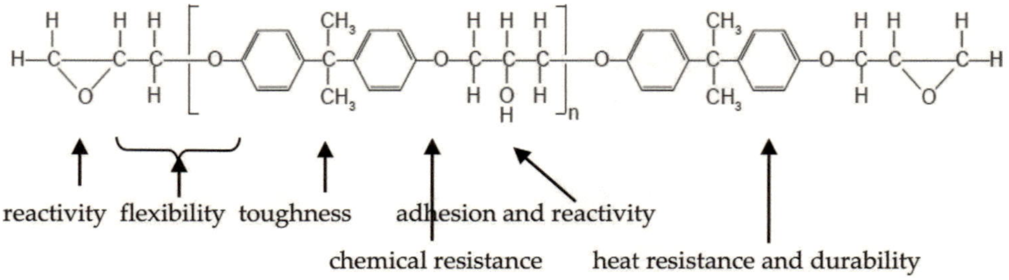

Figure 2.
Chemical structure and key features of DGEBA.

one-component package depending on the curing agent and curing method used. Two-component epoxy adhesive is prepared by packing epoxy composition and curing agent composition separately before use. It will cure soon after mixing based on designed mixing ratio. Almost all room temperature cure epoxy adhesives

Polymerization mechanism	Curing agent, initiator
Polyaddition	Polyamines Modified polyamines Mercaptans Phenols Anhydrates Dicyandiamide
Anionic	Tertiary amines Imidazole compounds
Cationic	Boron trifluoride monoethylamine Onium salts Iodonium salts

Table 4.
Epoxy curing agents.

are supplied in two-component package. Epoxy adhesives can be formulated in one-component package where all components including epoxy resin and curing agent has been mixed in advance. One-component epoxy adhesives usually use elevated temperature cure and need to be stored at low temperature conditions in a refrigerator or even freezer for long shelf life.

3. Room temperature cure epoxy adhesives

Room temperature cure epoxy adhesives are normally prepared and supplied in two-component package with epoxy resin component parked in one resin part and curing agent packed as the other hardener part. By mixing these two parts together, epoxy resin will react with curing agent quickly at room temperature conditions to become cross-linked strong thermoset structure that can bond adhesion substrates tightly. By use of different type of curing agents, pot life and cure time can be designed as needed.

3.1 Fast room temperature cure epoxy adhesives

Mercaptan compounds are usually selected as curing agent for fast room temperature curable epoxy adhesive because its reaction with epoxy resin is very fast in the existence of small amount of basic chemicals such as tertiary amine or imidazole as accelerator. As shown in **Figure 3**, epoxy resin reacts with mercaptan group equivalently via polyaddition reaction mechanism [14]. Fixture time can be <30 minutes or even 15 minutes at room temperature. Full cure time will need 24 hours. Precautions need to be paid on its very short work life, <10 minutes or even 5 minutes. Commercial fast cure epoxy adhesives supplied by Henkel AG & Co. KGaA and their typical properties are illustrated in **Table 5** [15].

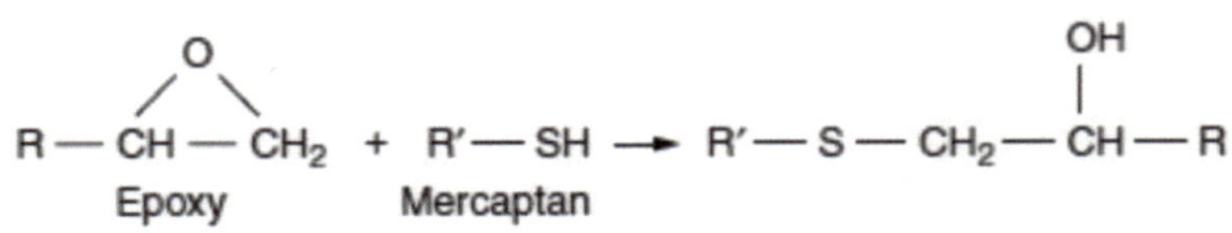

Figure 3.
Polyaddition reaction of epoxy resin with mercaptan.

Product	LOCTITE EA E-05MR	LOCTITE EA E-00NS
Color	clear	translucent
Viscosity, mPas/25°C	25,000	100,000
Mix ratio	1:1	1:1
Work life, minutes	5	3
Fixture time, minutes	15	10
Room temperature cure time, hours	24	24
Shear strength, psi on steel	3360	1600

Table 5.
Fast room temperature cure epoxy adhesives.

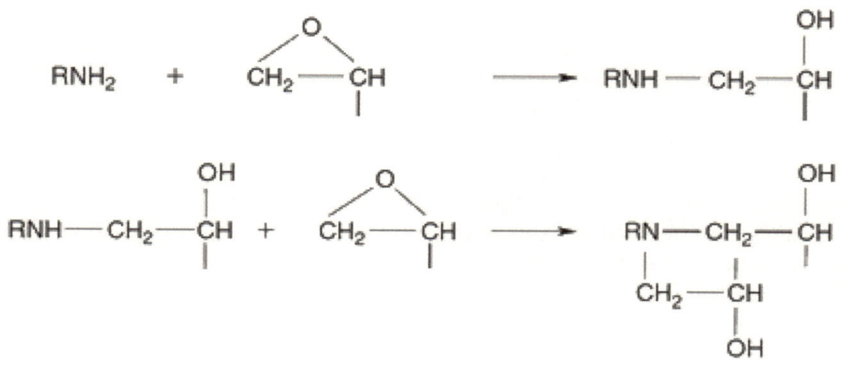

Figure 4.
Polyaddition reaction between epoxy resin and amine.

Product	LOCTITE EA E-20HP	LOCTITE EA E-120HP
Color	Off-White	Amber
Viscosity, mPas/25°C	30,000	30,000
Mix ratio	2:1	2:1
Work life, minutes	20	120
Fixture time, minutes	120	>180
Room temperature cure time, hours	24	24
Shear strength, psi on steel	3270	4300

Table 6.
Room temperature cure epoxy adhesives.

3.2 Room temperature cure epoxy adhesives

Aliphatic polyamines are most commonly used curing agents in epoxy resin technology. A number of modified polyamine type curing agents with adjustment on curability, handling or other physical properties for easy use have been commercialized in the market by curing agent suppliers. As shown in **Figure 4**, active hydrogen of primary and secondary amine reacts equivalently with epoxide via polyaddition mechanism [16]. Fixture time and work life can be adjusted by combination with suitable curing agent. **Table 6** shows commercial room temperature epoxy adhesives supplied by Henkel AG & Co. KGaA and their properties [17].

4. Thermal cure epoxy adhesives

Thermal cure epoxy adhesives are prepared and supplied in both one-component and two-component packages depending mainly on curing agent type used. Compared to room temperature cure type, thermal cure two-component epoxy adhesives usually have higher glass transition temperature that is suitable for high temperature resistance applications. One-component epoxy adhesives do not need pre-mixing in use and thus can be handled much easily. Many new one-component epoxy adhesives have been commercialized and become more and more important in recent years.

4.1 Two-component thermal cure epoxy adhesives

When use cycloaliphatic amine or aromatic amine as curing agent, post thermal cure process is usually required to achieve full cure as their reactivity, especially aromatic amines and secondary amine in cycloaliphatic amine, with epoxide is much lower with compared to aliphatic amines applicable for room temperature cure. Chemical structure of commonly used cycloaliphatic amine IPDA (isophorone diamine) and aromatic amine DDM (methylene dianiline) is shown in **Figure 5** [18, 19]. Thermal cure epoxy adhesives have much stronger and rigid structure and normally possess higher glass transition temperature with compared to room temperature cure epoxy adhesives mainly based on aliphatic amines or mercaptans. Two-component thermal cure epoxy adhesives are mainly used for higher temperature resistance required applications such as automobile production and aerospace market.

4.2 One-component thermal cure epoxy adhesives

One-component epoxy adhesives do not require pre-mix before use since all components have been mixed together and there is no concern on insufficient mixing problem as often occurred in two-component use. Pot life of one-component epoxy adhesives is usually long and one-component adhesives are thus suitable for automatic dispensing systems. Compared to two-component type, one-component epoxy adhesives can be handled much easily. On the other hand, one-component epoxy adhesives usually need cure at higher temperature because of long enough room temperature stability needed for adhesive preparation and storage. Most one-component epoxy adhesives require storage condition at lower temperatures in a refrigerator or even freezer.

Recently one-component thermal cure epoxy adhesives have become more and more important especially in electronics assembly and automotive production where high production efficiency is required. With selection of suitable latent curing agents, a number of one-component epoxy adhesives have been developed and commercialized by epoxy adhesive suppliers for various applications. Typical commercial latent curing agents are summarized in **Table 7**.

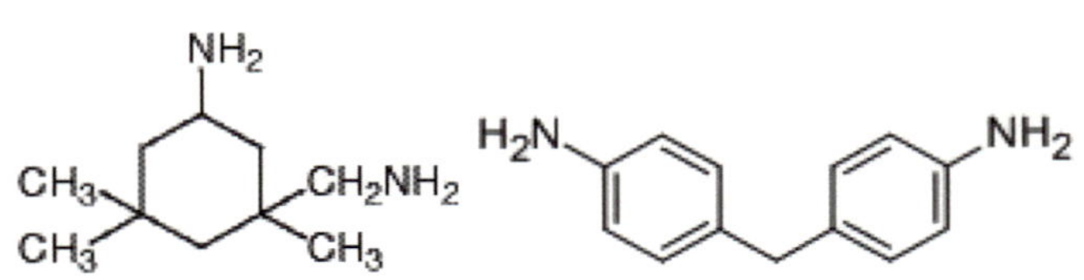

Figure 5.
Chemical structure of IPDA and DDM.

Latency mechanism	Latent curing agent	Curing agent state	Typical curing temperature °C
Chemical block and physical separation	DICY	Solid	≥150
	Dihydrazines		≥120
Physical separation	Modified imidazoles	Fine powder	≥80
	Modified polyamine		≥80
Chemical block	Onium salts	Solid	≥80
	Amine-BF3 complex	Liquid	≥130

Table 7.
Typical commercial latent curing agents.

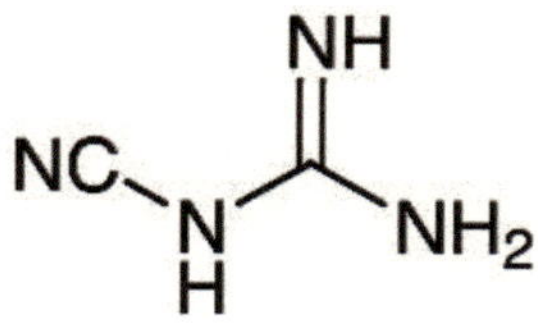

Figure 6.
Chemical structure of DICY.

DICY (dicyandiamide), chemical structure shown in **Figure 6** [20], is the oldest and widely used latent curing agent for epoxy resin technology. It is a solid chemical with a melting point at 208°C. DICY formulated epoxy composition is very stable, up to 6 months at room temperature. Latency mechanism is a combination of physically separation and chemically blocking with epoxide group. DICY cured epoxy resin shows high adhesion and possesses high glass transition temperature especially suitable for high performance required applications such as vehicle parts bonding in automobile production. Cure temperature of DICY alone with epoxy resin normally needs at least 150°C to achieve full cure. By addition small amount of accelerator such as modified urea compounds and imidazole compound, cure temperature can be further lowered to 120°C [21].

In recent years, new type latent curing agents have been developed and commercialized by several curing agent suppliers [22–26]. These latent curing agents are supplied as fine powder with average particle size well controlled in a few microns or premix of fine powder latent curing agent in epoxy resin. They are manufactured by grinding specially synthesized modified polyamine or imidazole solid with a softening point from 80 to 150°C. Latency mechanism is mainly physically separation between curing agent and epoxide. Curing temperature has been be lowered to as low as 80°C and its formulated epoxy composition can be still quite stable at room temperature. Many one-component epoxy adhesives commercialized recently are based on these new type latent curing agents because of their lower temperature curability suitable for use in bonding heat sensitive substrates such as plastics. By combination of small amount of liquid phenol compound with latent curing agent, it has been found that cure time of one-component epoxy adhesives can be shortened significantly [27, 28].

5. UV cure epoxy adhesives

Ultra-violet light (UV) curable epoxy adhesives can be cured quickly and have been very successfully used in several new electronics assembly and general bonding

Adhesive type	UV acrylate	UV cationic epoxy	Hybrid thermal cure epoxy
Main compositions	Acrylate	Epoxy resin	Acrylate
	Photoinitiator	Cationic photoinitiator	Photoinitiator
			Epoxy resin
			Curing agent
Polymerization			
UV cure	Radical	Cationic	Radical
Thermal cure	N.A.	Cationic	Polyaddition, anionic
Key features			
Oxygen inhibition	Yes	Yes	Partially
Alkali inhibition	No	Medium	No
UV curability	High	Preferred	High
Post thermal cure	No need	Partially	Need
Shadow cure	No	Low	Yes
Cure shrinkage	High	Good	Low
Adhesion	Moderate		Good

Table 8.
Comparison of UV acrylate, cationic epoxy and hybrid epoxy adhesives.

applications such as image sensor module assembly, display panel and module assembly where fast production speed and high adhesion performance are required. Various UV cationic epoxy adhesive and UV acrylate hybrid thermal cure epoxy adhesives have been commercialized in recent years. As compared in **Table 8**, UV cure epoxy adhesives have no oxygen inhibition issue, low curing shrinkage and show better adhesion with compared to common UV acrylate adhesives.

5.1 UV cationic epoxy adhesives

UV cationic epoxy adhesives are primarily composed of epoxy resin and cationic photo-initiator [29–31]. Cycloaliphatic type epoxy resins are usually selected for UV cationic epoxy adhesives because of faster cationic polymerization rate than that of normal bisphenol A diglycidyl ether type epoxy resin. As illustrated in **Figure 7** [32], cationic photoinitiator formulated in UV epoxy adhesives absorbs UV energy to generate strong acid that will react with epoxy to produce cationic which can initiate homo-polymerization of epoxy resin. Compared to common acrylate based UV adhesives, UV cationic epoxy adhesives have lower cure shrinkage because of epoxy structure and have no surface cure issue resulted from oxygen inhibition to free radical polymerization since they cure via cationic polymerization. On the other hand, UV cationic epoxy adhesives are not suitable for bonding basic substrates which terminate cationic polymerization. UV cationic epoxy adhesives will need some longer cure time. In real use, a post thermal cure of UV cationic epoxy adhesives after the UV radiation is commonly used for full cure to assure satisfactory adhesion performance.

UV cationic epoxy adhesives have been commercialized and used in optical parts bonding, camera module sensor packaging and OLED panel assembly applications [33–37]. The authors have found that adhesion reliability performance of UV cationic epoxy adhesives can be much improved by combination use of cationic photo initiator with thermal cationic initiator [38].

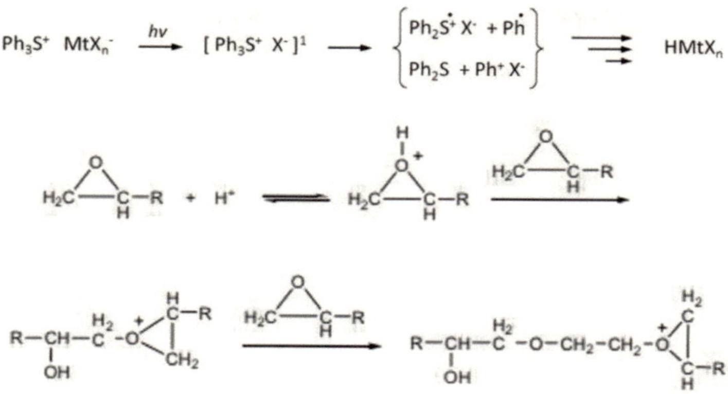

Figure 7.
UV cationic polymerization of epoxy adhesives.

5.2 UV hybrid epoxy adhesives

Most widely used UV cure adhesives are acrylate compositions [39–41]. Acrylate based UV cure adhesives are mainly composed of acrylate monomer, acrylate oligomer and radical photo-initiator. Acrylate based UV cure adhesives can be cured instantly, within seconds. Limitations of UV cure acrylate adhesives are surface cure issue, shadow cure problem, high cure shrinkage and poor humidity reliability. By combination of UV acrylate composition with thermal cure epoxy composition, UV and thermal cure hybrid epoxy adhesives have been developed and commercialized for over two decades [42–44]. Acrylate monomer, epoxy resin, photo-initiator and epoxy curing agent are primarily formulated in the UV and thermal cure hybrid adhesives. UV hybrid epoxy adhesives combine advantages from both UV acrylate proportion and thermal cure epoxy part. Adhesion reliability performance could be much improved by introduction of epoxy composition with compared to normal UV acrylate adhesives. In the meantime, production efficiency could be much improved by shortening the fixture time to seconds via UV radiation with compared to at least dozens of minutes needed for thermal cure epoxy adhesives. Surface cure issue, shadow cure issue and cure shrinkage problem of acrylate UV adhesives could also be improved to certain degree because of lower contents of free radical curable acrylate compositions. In some cases, thermal initiator such as peroxide is also formulated in the hybrid adhesive to assure curing remained acrylate compositions after UV radiation or those at shadow areas where UV light cannot reach.

Successful development and industrialization of so called ODF (One Drop Fill) process for large size LCD (liquid crystal panel) panel production was one important technology revolution in early 2000s that have made a big impact on our modern life. Development and commercialization of LCD ODF main sealant, an UV hybrid epoxy adhesive, played a key role in its mass production success [45–47]. LCD ODF main sealant is an adhesive material that is used to bond two glass substrates and seal liquid crystal material between them. It is a UV hybrid epoxy adhesive, typically composed of acrylate monomer, photo-initiator, partially acrylate epoxy resin and latent curing agent. Main steps for the adhesive use in this process are: (1) dispensing LCD main sealant; (2) dropping off liquid crystal materials into each cell; (3) alignment and assembly; (4) UV cure the sealant; and (5) thermal cure the sealant. The author has invented initiator free UV hybrid thermal cure epoxy adhesives by combination with bismaleimides that shows much better compatibility with liquid crystal material and high performance [48–50].

6. Summary

Epoxy adhesives show very good adhesion to various substrates and are the most important structural adhesives. Epoxy adhesives can be cured at room temperature condition, at elevated temperature condition or via UV light radiation mainly depending on curing agent type formulated. Lots of epoxy adhesives, either supplied in one-component or two-component package, have been commercialized and widely used for bonding metals, concrete, glass, ceramics, concrete, many plastics, wood, etc. in various industrial production and applications.

Author details

Chunfu Chen[1]*, Bin Li[2], Masao Kanari[1] and Daoqiang Lu[2]

1 Henkel Technology Center-Asia Pacific, Henkel Japan Ltd., Yokohama, Japan

2 Henkel Adhesive Innovation Center, Henkel (China) Co., Ltd., Shanghai, PR China

*Address all correspondence to: chunfu.chen@henkel.com

References

[1] Petrie EM. Handbook of Adhesives and Sealants. New York: McGraw-Hill; 2006. 355 p.

[2] Ha QP, Marks MJ. Epoxy resins. In: Ley C, editor. Encyclopedia of Polymer Science and Technology. New Jersey: Wiley; 2004. DOI: 10.1002/0471440264.pst119

[3] Panda H. Epoxy Resins Technology Handbook. New Delhi: Asia Pacific Business Press; 2016

[4] Sancaktar E, Bai L. Electrically conductive epoxy adhesives. Polymers. 2011;**3**:427-466. DOI: 10.3990/polym3010427

[5] Severijin C, Teixeira de Freitas S, Poulis JA. Susceptor-assisted induction curing behavior of a two-component epoxy paste adhesive for aerospace applications. International Journal of Adhesion and Adhesives. 2017;**75**:155-164. DOI: 10.1016/j.ijadhadh.2017.03.005

[6] Vidil T, Tournilhac F, Musso S, Robisson A, Leibler L. Control of reactions and network structures of epoxy thermosets. Progress in Polymer Science. 2016;**62**:126-179. DOI: 10.1016/j.progpolymsci.2016.06.03

[7] Zotti A, Zuppolini S, Zarrelli M, Borriello A. Fracture toughening mechanisms in epoxy adhesives. In: Adheisves—Applications and Properties. London: IntechOpen; 2016. pp. 237-269. DOI: 10.57772/65250

[8] Lewis AF. Epoxy resin adhesives. In: May CA, editor. Epoxy Resins—Chemistry and Technology. 2nd ed. New York: Marcel Dekker; 1988. 653 p

[9] Jin F-L, Li X, Park S-J. Synthesis and applications of epoxy resin: A review. Journal of Industry and Engineering Chemistry. 2015;**29**:1-11. DOI: 10.1016/j.jiec.2015.03.026

[10] Groulding TM. Epoxy resin adhesives. In: Pizzi A, Mittal KL, editors. Handbook of Adhesive Technology. 2nd ed. New York: Marcel Dekker; 2003. pp. 809-824

[11] Petrie EM. Epoxy Adhesive Formulations. New York: McGraw-Hill; 2006. DOI: 10.1036/0071455442

[12] Liu JQ, Bai C, Jia DD, Liu WL, He FY, Liu QZ, et al. Design and fabrication of a novel superhydrophobic surface based on a copolymer of styrene and bisphenol A diglycidyl ether monoacrylate. RSC Advances. 2014;**4**:18025-18032. DOI: 10.1039/4cra01505c

[13] Muroi S, Ishimura H. Epoxy Resin Introduction. Polymer Publishing Association; 1988. p. 2

[14] Fernandez-Frncos X, Konuray AO, Belmonte A, De la Flor S, Serra A, Ramis X. Sequential curing of off-stoichiometric thiol-epoxy thermosets with a custom-tailored structure. Polymer Chemistry. 2016:2280-2290. DOI: 10.1039/c6py00099a

[15] Technical Data Sheet and Product Catalog of LOCTITE EA E-05MR and EA E-00NS. Henkel Corporation; 2015

[16] Thomas R, Sinturel C, Thomas S, El Akiaby EMS. Introduction. In: Thomas S, Sinturel C, Thomas R, editors. Micro- and Nanostructured Epoxy/Rubber Blends. Weinheim: Wiley-VCH Verlag; 2014. 3 p.

[17] Technical Data Sheet and Product Catalog of LOCTITE EA E-20HP and EA E-120HP. Henkel Corporation. 2015

[18] Sigma-Aldrich (Merck KGaA). Isophorone diamine. Available from: https://www.sigmaaldrich.com/catalog/product/aldrich/118184 [Accessed: 2019-04-01]

[19] Sigma-Aldrich (Merck KGaA). Methylene diamine. Available from:

https://www.sigmaaldrich.com/catalog/product/aldrich/32950 [Accessed: 2019-04-01]

[20] Tokyo Chemical Industry Co., Ltd. Dicyandiamide. Available from: https://www.tcichemicals.com/eshop/ja/commodity/c0454/ [Accessed: 2019-04-01]

[21] Guthner T, Hammer B. Curing of epoxy resins with dicyadiamide and urones. Journal of Applied Polymer Science. 1993:1453-1459. DOI: 10.1002/app.1993.070500817

[22] Ajinomoto Fine-Techno Co., Ltd. Latent curing agent "AJICURE". Available from: https://www.aft-website.com/en/chemistry/ajicure [Accessed: 2019-04-01]

[23] ADEKA Corporation. ADEKA Hardener EH Series (Latent Hardener). Available from: https://www.adeka.co.jp/en/chemical/products/functional/pro143c.html#sp05 [Accessed: 2019-04-01]

[24] Evonik Corporation. Epoxy Additives and Polyamides. Available from: https://crosslinkers.evonik.com/product/crosslinkers/downloads/evonik-epoxyguide_asiapacific.pdf [Accessed: 2019-04-01]

[25] Asahi Kasei Advance Corporation. Novacure. Available from: https://www.asahi-kasei.co.jp/advance/en/business/plastics/plastics04.html [Accessed: 2019-04-01]

[26] T & K TOKA Corporation. Latent Hardener "FUJICURE". Available from: https://www.tk-toka.co.jp/product/jushi/epoxy/detail/files/fujicure_solid.pdf [Accessed: 2019-04-01]

[27] Chen C. Thermal curable liquid resin composition. Japan Patent 4204814

[28] Chen C. One component epoxy resin composition. European Patent 2640765

[29] Yu V, Voytekunas FLN, Marc JMA. Kinetics study of the UV-initiated cationic polymerization of cycloaliphatic diepoxide resins. European Polymer Journal. 2008;**44**:3640-3649. DOI: 10.106/j.eurpolymj.2008.08.043

[30] Golaz B, Michaud V, Leterrie Y, Manson J-AE. UV intensity, temperature and dark-curing effects in cationic photo-polymerization of a cycloaliphatic epoxy resin. Polymer. 2012;**53**:2038-2048. DOI: 10.1016/j.polymer.2012.03.025

[31] Atif M, Bongiovanni R, Yang J. Cationically UV-cured epoxy composites. Polymer Reviews. 2015;**55**:90-106. DOI: 10.1080/15583724.2014.963236

[32] Corcione C, Malucelli G, Frigione M, Maffezzoli A. UV-curable epoxy systems containing hyperbranched polymers: Kinetics investigation by photo-DSC and real-time FT-IR experiments. Polymer Testing. 2009;**28**:157-164. DOI: 10.1016/j.polymertesting.2008.11.002

[33] Chiang TH, Hsieh TE. A study of monomer's effect on adhesion strength of UV-curable resins. International Journal of Adhesion and Adhesives. 2006;**26**:520-531. DOI: 10.1016/j.ijadhadh.2005.07.004

[34] Gan Y, Chen C, Terada K. Cationically photocurable epoxy resin composition. US Patent 7456230

[35] Velankar S, Pazos J, Cooper SL. High-performance UV-curable urethane acrylates via deblocking chemistry. Journal of Applied Polymer Science. 1996;**62**:1361-1376

[36] Fourassier JP, Lalevee J. Photoiniator for Polymer Synthesis. Weinheim: Wiley-VCH Verlag; 2012. 41p

[37] Ebnesajjad S. Adhesive Technology Handbook. 2nd ed. London: William Andew; 2008. 124p

[38] Chen C, Gan Y. Cationically curable epoxy resin composition. US Patent 7795744

[39] Vitale A, Trusiano G, Bongiovanni R. UV-curing of adhesives: A ctitical review. In: Mittal KL, editor. Progression in Adhesion and Adhesive, III. Beverly: Scrivener Publishing; 2018. 101p. DOI: 10.1002/9781119526445.ch4

[40] Dekker C. UV-radiation curing of adhesives. In: Cognard P, editor. Adhesives and Sealants. New York: Elsevier; 2006. 303p

[41] Goss B. Bonding glass and other substrates with UV curing adhesives. International Journal of Adhesion and Adhesives. 2002;**22**:405-408

[42] Park C, Lee S, Park J, Kim H. Preparation and characterization of dual curable adhesives containing epoxy and acrylate functionalities. Reactive & Functional Polymers. 2013;**73**:641-646. DOI: 10.1016/j.reatfunctpolym.2013.01.012

[43] Xiao M, He Y, Nie J. Novel bisphenol A epoxide-acrylate hybrid oligomer and its photopolymerization. Designed Monomers and Polymers. 2008;**11**: 383-394. DOI: 10.1163/156855508X332522

[44] Park Y, Lim D, Kim H, Park D, Sung I. UV- and thermal-curing behavior of dual-curable adhesives based on epoxy acrylate oligomers. International Journal of Adhesion and Adhesives. 2009;**29**:710-717. DOI: 10.1016/j.ijadhadh.2009.02.001

[45] Matsuda M. Sealants for one drop fill (ODF) process. In: Koide N, editor. The Liquid Crystal Display Story. Tokyo: Springer; 2014. 199p

[46] Hirai A, Abe I, Mitsumoto M, Ishida S. One drop filling for liquid crystal display panel produced from large size mother glass. Hitachi Review. 2008:144-148

[47] Chen C, Iida K. Adhesive for flat-panel display manufacture. In: Ullmann's Encyclopedia of Industrial Chemistry. Weinheim: Wiley-VCH; 2010. DOI: 10.1002/14356007.n01

[48] Chen CF, Iwasaki S, Kanari M, Li B, Wang C, Lu D. High performance UV and thermal cure hybrid epoxy adhesive. IOP Conference Series: Materials Science and Engineering. 2017;**213**:012032. DOI: 10.1088/1757-899X/213/1/012032

[49] Chen C. Sealing agent for liquid crystal dropping technology and method of manufacturing liquid crystal display. Japan Patent 5592081

[50] Chen C. Sealant composition. US Patent 10108029

Chapter 4

Overhauling of Steel Pipes Using Vacuum Bagging Processed CFRP Patch

Syam Kumar Chokka, Beera Satish Ben and Kowtha Venkata Sai Srinadh

Abstract

The present work deals with the experimental study on rehabilitation of damaged steel pipes. The process of rehabilitation was done by using adhesive bonded CFRP patch on damaged sites of the pipes with the help of vacuum bagging set-up. The reference optimal parameters to rehabilitate the damaged pipes were considered from the tensile test (tensile shear load) conducted on plates. The rehabilitated pipes were tested under hydrostatic pressure. The two-component structural adhesive (Araldite AW105 and Hardener HV 953U) used in this study has high viscosity. It is not desirable for CFRP composite making and may lead to improper penetration of adhesive through open pores of the adherent surface. The viscosity of the adhesive was reduced in two ways: addition of a low viscous two-component nonstructural adhesive (Araldite LY-1564 and Aradur-22962) and heating the adhesive. Vacuum pressure, bond length and pre-bond surface preparation of the adherents were considered as parameters to evaluate the bond strength. The roughness of different samples was studied using a 3D microscope. The surface morphology of adherent was studied using scanning electron microscopy (SEM). Based on the experimental studies, it is observed that the optimal conditions of the tensile data of the plates hold good for the rehabilitated pipe under hydrostatic loading.

Keywords: vacuum bagging process, adhesive viscosity, surface pre-bond preparation, carbon fibre, SS-304 plates and pipes

1. Introduction

Fibre-reinforced polymer (FRP) composites are made of a polymer matrix reinforced with fibres. They are different types of fibres generally carbon, glass, aramid fibres, etc., while the polymers are generally an epoxy, polyesters, vinyl esters, etc. Polymer composites are used in automotive, marine, construction and aerospace industries. Carbon fibre-reinforced polymer (CFRP) composites have high strength to weight ratio and are resistant to most chemical attacks.

There is a huge demand for repairing or strengthening of existing structures in many applications either to sustain increased loads or to repair deteriorated structures. Many structural members like metallic pipelines, bridges and such other structures may deteriorate as they age, where the conventional method of repairing involves replacement or welding another plate over the damaged plate. The plates used to rehabilitate the structure are usually heavy, the process is difficult and chances of corrosion and fatigue are high as they are subjected to high temperatures while welding. Adhesive bonded CFRP patch can be a good replacement for conventional methods as it has a high strength to weight ratio with design flexibility [1, 2]. Adhesive bonding can be done for similar, dissimilar and thin components. There are many applications in marine structures for adhesive bonding such as bonding of the hull to the deck, the channels running through the deck, the sea chests, engine compartment and the exhaust system. Repairing with CFRP patch can give a long life to the components as they have high resistance to chemical attacks and corrosion.

The cost of the structural repairing with adhesive bond using wet lay-up method can be reduced [3, 4]. In this process, fibre mats were wetted by an adhesive/resin until the desired thickness is achieved. Rollers are used to promote good impregnation of the fibre and reduce the voids entrapped in the adhesive mixture. The use of wet/hand lay-up results from the ease of application, minimal tooling cost and low cost of raw material. But the emissions from the adhesive/resin, low fibre volume fraction and high void content may limit its use [5–8]. The hand/wet lay-up process along with vacuum bag can prevent the harmful emissions and produce better quality products with the help of vacuum pressure.

Vacuum bagging process is used in many repairing applications in construction, marine and aerospace industries. Generally, scarf repair uses wet lay-up vacuum bagging process to bond CFRP patch on the surface of the damaged components. However, generally, the bond between the adherents is weaker than adherents in most cases due to the inability to apply sufficiently high-pressure during curing process. Thus, piles are added to the patch to attain the desired stiffness and load-carrying capacity [9]. The increased thickness of the patch enhances the bending stiffness and leads to premature failure at the edges under bending loads [10]. The properties of the repair patch can be enhanced by reducing the void content and increasing the fibre volume fraction which may be possible by vacuum bagging process.

Adhesive bonding can be done for similar, dissimilar and thin components. But the problem with adhesive bonding of metal plates with CFRP is its low strength. The major adhesion mechanisms that occur between CFRP and metals are (a) chemical bonding such as van der Waals forces, (b) mechanical interlocking between adhesive and substrate and (c) diffusion bonding. Joint strength for interdiffusion phenomena depends on diverse aspects, namely, contact temperature, time, nature and molecular weight of the polymers, etc. In order to vary the mechanical interlocking of surfaces, pre-bond surface preparation is needed, which includes surface abrasion, sand blasting, etching, etc. [11]. Mechanical interlocking is provided by allowing adhesive to wet the cavities and asperities of adherent surface. However, the surface asperity dimension should be controlled to avoid the formation of air bubbles. Air bubbles can generate regions of stress concentrations which are not desirable. In most cases, rough surface is good for better mechanical interlocking between the adhesive and adherent.

The evaporation of resin is another source of void formation in composites. The evaporation was high at the beginning of curing at room temperature. After 30 min, the evaporation settles to a constant rate. The high evaporation speed at the beginning was probably caused by air dissolving in the resin. The air bubbles can be removed with the application of vacuum to the mould [12, 13]. The application of vacuum pressure may pull the air bubbles formed during the process out of the mould. If the vacuum pressure is too high, the carbon fibre preform arrests the air bubbles formed between the laminates and creates voids [14]. The increased vacuum pressure may increase the compaction pressure on the carbon fibre preform which in turn reduces the resin content in the final product.

To rehabilitate the damaged SS pipe using a composite patch (CFRP) prepared with vacuum bagging process, the important aspect to study is the bond interface between SS and CFRP [15]. Single-strap adhesive joint is the least strength joint configuration than any others [16]. The present work aims to enhance the adhesive bond strength (tensile shear load capacity) of single-strap SS-CFRP joint using vacuum bagging process. The parameters for investigating the effect on load capacity of the joint are adhesive combination and temperature, vacuum pressure, bond length and surface of the adherent with different surface textures. These optimal parameters are applied to rehabilitate the damaged SS pipes and investigate its hydrostatic pressure resistance.

2. Materials and methods

2.1 Materials

Major materials used for the current study along with their properties were given in **Table 1**. The properties of the adhesives were taken from the Huntsman Advanced Materials, Switzerland data sheet. Fibre properties were considered form the Toray Composite Materials America, Inc. data sheet. Along with major materials, consumable like ferric chloride, distilled water, sticker tape, vacuum bagging, etc. were also used.

S. no	Materials		Viscosity [Pa s]	Tensile strength [MPa]	Elastic modulus [GPa]	Density at 25°C [g/cc]
1	Structural adhesive (10:8 mix ratio)	Araldite-AW 106	30–50	Not given	Not given	1.14–1.15
		Hardener-HV 953 U	20–35			0.94–0.95
2	Nonstructural adhesive (10:2.5 mix ratio)	Araldite LY 1564	1.2–1.4	75–80	2.8–3.3	1.10–1.20
		Aradur-22962	0.005–0.02			0.89–0.90
3	Fibre	12KUD-300 Gsm UD-carbon fibre	NA	3500	230	1.76
4	Stainless steel	SS 304	NA	550	200	7.85

Table 1.
Materials and properties.

2.2 SS plate preparation

SS plates were considered in the following dimensions: 125 × 25 × 3 mm. The surface of the samples was prepared with different surface preparation methods. The surface preparation includes surface without preparation cleaned with acetone (plane surface), surface prepared with etching, surface prepared with sand blasting and the sandblasted surface with open surface cavities.

The open surface cavities were made on the SS plates using chemical etching process. To perform chemical etching on the samples, the part of material to be etched should be opened to the chemical interaction, and the rest of the material was masked with sticker tape. Ferric chloride and distilled water were mixed in the ration of 1.5:10. The masked samples were dipped and rinsed in the etchant (ferric chloride and distilled water mixer) for 8 min to get 80 ± 0.1-micron deep surface cavities. The etching process and the etched 3D microscopic image were shown in **Figure 1a**. The pipe surface was also prepared with plane surface cleaned with acetone, sandblasted and chemical etched and a combination of sandblasting and surface cavities.

The plates prepared with different methods show different surface morphologies. These morphologies can be seen using SEM images as shown in **Figure 2a–c**. From SEM images it was evident that the roughness of chemically etched surface is lesser than the roughness of the sandblasted surface. There are some locations on the sandblasted surface where the sand particulates diffused onto the surface of the specimens.

2.3 Single-strap SS-CFRP adhesive joint preparation

A flat marble was considered as a mould. The boundaries of the moulds were marked, and the sealant tape was attached. The pre-bond surface prepared samples were cleaned with acetone to remove chemical residues, dust and contaminations from the surface. The gap between the plates should be less than 1 mm in order to avoid the edge effect. Alternate layers of adhesive and carbon fibre were applied onto the surface of the plates (3 layers of CFRP and 4 layers of adhesive). Different accessory layers were placed over the carbon fibre layers to assist the vacuum bagging process. The accessory layers include a peel ply which is used to provide easy removal of other layers from the surface of the composite and a breather fabric to absorb excess resin during compression (vacuum pressure). Vacuum bag was applied with the help of the sealant tape, which forms a one-sided flexible mould. The mould was connected to the vacuum reservoir through vacuum hose pipe as shown in **Figure 3a** and **b**. The vacuum bag should be checked for vacuum drop (leakage). To do so the hose pipe should be closed, while the vacuum bag should be connected the vacuum gauge. A good vacuum bag should not drop more than 500 Pa in 5 min, and it is not recommended to be used if the vacuum loss is greater than 5000 Pa in 5 min. The mould should be isolated from the vacuum reservoir through proper clamping devices and allow the samples to cure for 24 hrs in the atmospheric condition. When demoulding, care must be taken as the excess adhesive would spread and stick over the entire mould cavity. Machining was done to remove the excess adhesive. The schematic of sample dimensions can be seen in **Figure 3c**.

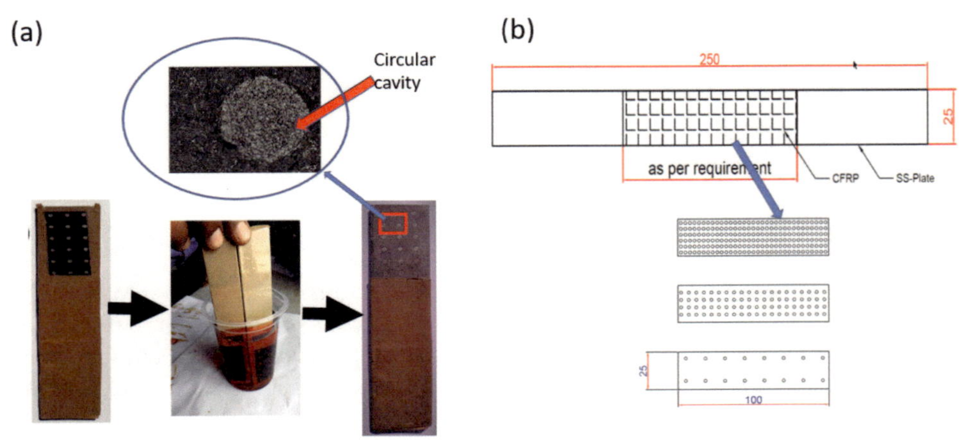

Figure 1.
(a) Chemical etching process and (b) cavities with different densities.

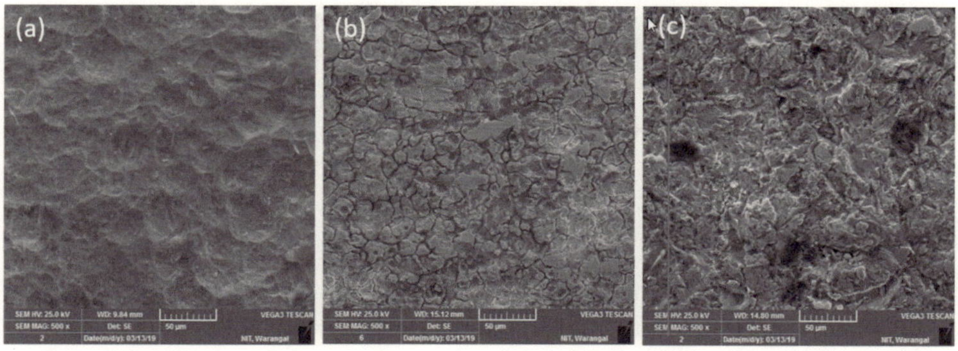

Figure 2.
SEM images (a) chemical etched surface; (b) plane surface; (c) sandblasted surface.

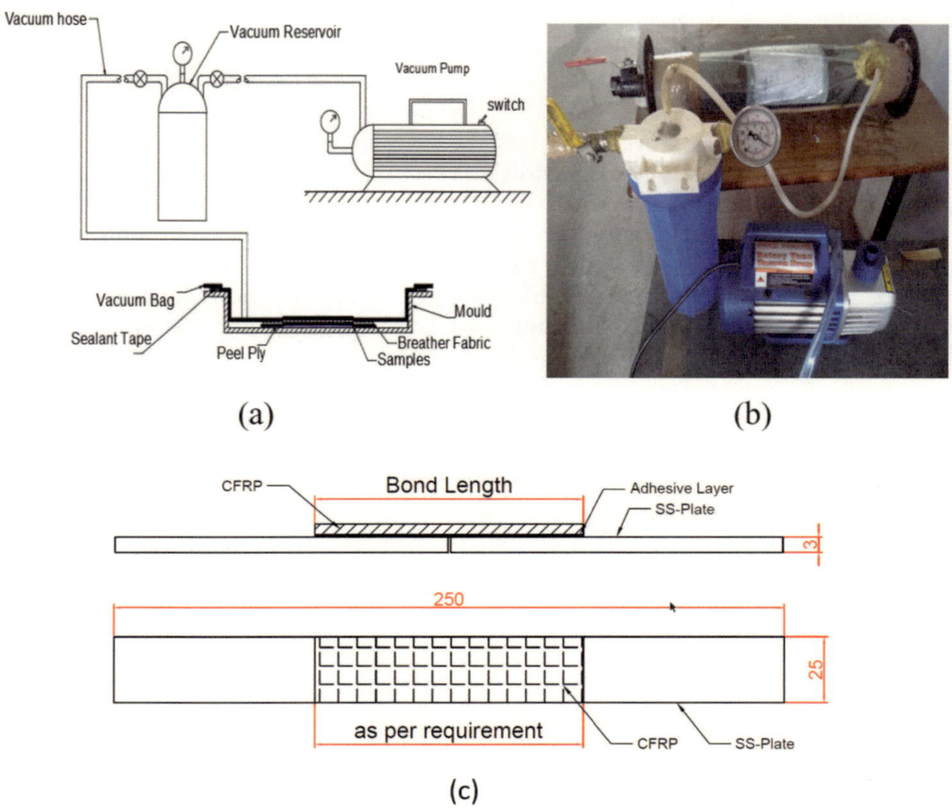

Figure 3.
(a) Schematic of vacuum infusion set-up; (b) homemade vacuum bagging set-up; (c) sample dimensions (mm).

2.4 Characterisation

The adhesive bond strength of SS plates was evaluated with tensile shear load capacity. It was measured using M/s. Jin Ahn Testing, China, with 100 kN load cell under a crosshead rate of 0.5 mm/min. The results shown here are the average values of three samples. The roughness of different pre-bond prepared surfaces was measured using 3D optical microscope (Huvitz Automatic 3D Measuring

Microscope). The surface morphology of the prepared samples was analysed using SEM micrographs taken with the help of "TESCAN VEGA3 LMU SEM". All tests were performed at room temperature except viscosity measurement.

3. Results and discussion

3.1 Adhesive viscosity

The two-component structural adhesive (Araldite AW105 and Hardener HV 953U) used in this study has high viscosity. It is not desirable for CFRP composite making and penetration of adhesive through open pores of the surface of the adherents. The adhesive viscosity was reduced in two ways: (1) addition of low viscous nonstructural adhesive and (2) heating the adhesive to get required viscosity.

3.1.1 Addition of low viscous nonstructural adhesive

The adhesive mixture was taken with different proportions by adding nonstructural adhesive in structural adhesive. The samples were prepared as per ASTM D638 standards as shown in **Figure 5c**, and tensile test was performed. The change in the viscosity of the mixture with NSA addition was calculated using Gambill method, and it was shown in **Figure 4a**.

Single-strap SS-CFRP adhesive bonded samples were prepared by adding a low viscous nonstructural adhesive (NSA) from 0 to 100% in structural adhesive (SA). The variable parameters like surface preparation, bond length and vacuum pressure were taken at random as sand blasted, 100 and 350 mm of Hg. The samples were cured in atmospheric conditions and de-moulded after 24 h of curing. The samples were machined to remove the excess material. The samples were tested in the UTM for tensile shear load capacity of the adhesive bonded joints prepared. The results were plotted between % NSA and tensile shear load capacity as shown in **Figure 4b**. The addition of NSA decreases the load capacity of the joint and the viscosity of the adhesive mixture [17]. Generally, NSA does not take any loads they meant for holding the fibres or components together, while the SA is capable of bearing loads. Hence the addition of NSA reduces the adhesion strength.

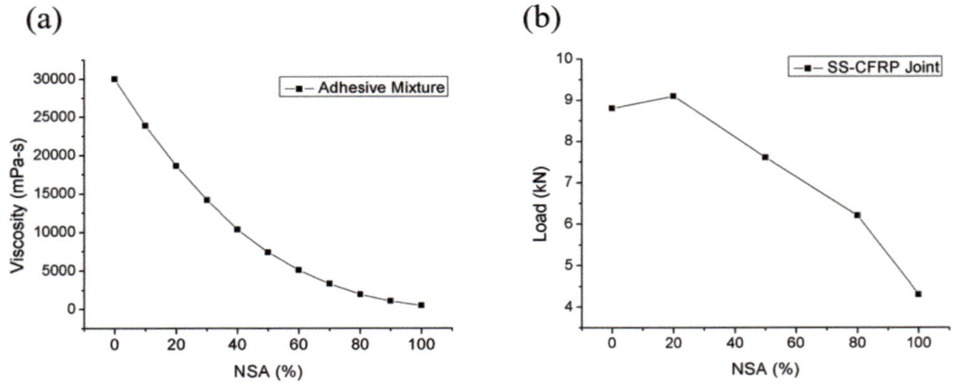

Figure 4.
Effect of NSA addition on (a) viscosity of the adhesive mixture; (b) load capacity of the SS-CCRP joint.

The adhesive joint with pure nonstructural adhesive has a load capacity of 4.3 kN. It reaches to a maximum of 9.1 kN with 20% NSA addition and 8.8 kN with pure structural adhesive. The viscosity with 20% NSA addition would be enough to penetrate through the open pores. Hence the mechanical interlocking and load capacity of the joint increases.

3.1.2 Heating effect

Curing of the adhesive makes the resin to form a cross-linked network of polymers. The structure and its mechanical properties can be changed by the way the chain network forms, which depends on the curing process. During solidification of adhesive, there is a shrinkage which creates some internal stresses in the adhesive and leads to failure of the joint well below its designed load [17]. An attempt has been made to find the effect of adhesive pre-cure temperature on tensile strength. The viscosity of adhesive was reduced by increasing the temperature of the adhesive. The adhesive was heated to 45, 55, 65, 72 and 80°C in a furnace. The viscosity at each temperature was measured using "Brookfield Viscometer". The results were plotted between temperature and viscosity as shown in **Figure 5a**. The increased curing temperature may also change the curing rate, which in turn affects the bond strength. A series of experiments were performed on adhesive samples cured at different temperatures (45, 55, 65, 72 and 80°C) to evaluate the curing temperature effect on tensile strength of the adhesive. The sample dimensions were considered as per ASTM D638 standards. The results obtained from the tensile test can be seen in **Figure 5b**.

From **Figure 5b** and **d**, it was observed that the tensile strength of the adhesive and tensile shear load capacity of the SS-CFRP joint were increased with the increase in precuring temperature [18]. The adhesive cured at lower temperatures (45°C) has shown a brittle nature than the adhesive cured at higher temperature (80°C) during tensile test.

From **Figure 5a** and **b**, it is evident that the adhesive has shown a viscosity of 560 mPa-s at 80°C and a tensile strength of 21.1 MPa which is optimal when considering both viscosity and tensile strength as a function of temperature. In case of adhesive, joint preheat temperature limits the use of adhesive preheat temperature to 55°C. Further heating reduces the gel time, and curing it might not be good for vacuum bagging process. The supplier's data states that the increased temperature reduces the minimum curing time ("At 20°C the minimum curing time is 15 hrs, whereas at 100°C it is 10 min"). Hence the adhesive curing temperature was considered as 55°C where the minimum curing time is about 2 hrs (gel time is directly proportional to curing time) which is enough to prepare the sample.

The addition of NSA in the adhesive mixture reduces the adhesive bond strength. Heating the adhesive mixture reduces the gel time. Hence the adhesive was considered as 20% NSA + 80% SA mixture heated to 55°C which has a viscosity of 950 mPa-s.

3.2 Effect of bond length

The stress generated at the edge is maximum in the joint. As the bond length increases, the stress generated at the edge reduces. But it is up to a certain length beyond which the addition of bond length has no significance. Then the increased load that may act at the edge undergoes an elastic-plastic transition [19].

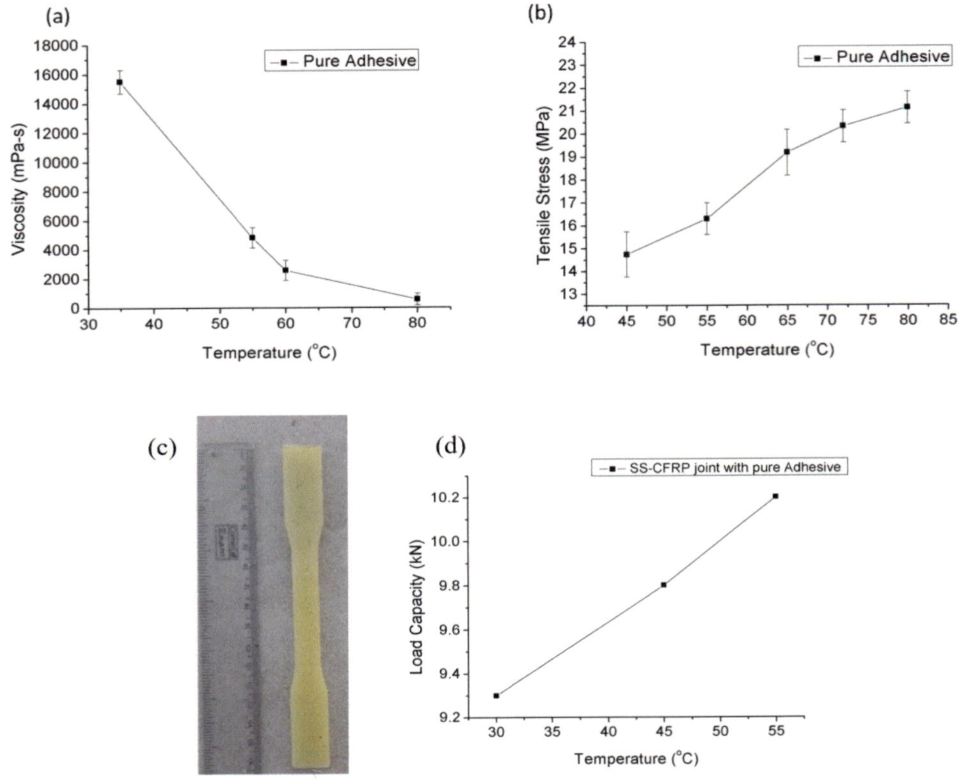

Figure 5.
(a) Viscosity of adhesive as a function of temperatures; (b) tensile strength of the adhesive as a function of temperatures; (c) adhesive tensile test sample; (d) SS-CFRP joint with different preheat temperatures.

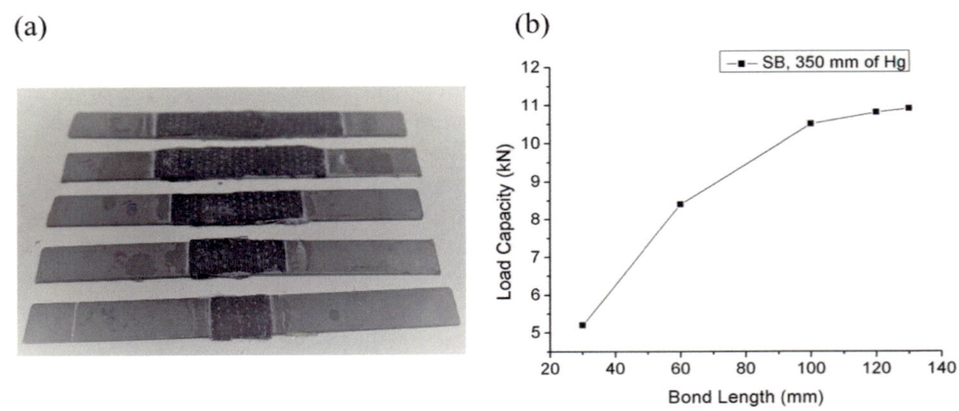

Figure 6.
(a) Samples with different bond lengths; (b) effect of bond length on load capacity.

To evaluate the effect of bond length on joint load capacity, the adhesive bonded samples were prepared with different bond lengths as shown in **Figure 6a**. The parameters considered for this study are 20% NSA + 80% SA mixture heated to 55°C, sandblasted surface and 350 mm of Hg vacuum pressure. The samples were tested for tensile load capacity, and the obtained results were plotted as shown in **Figure 6b**.

From **Figure 6b**, it has been observed that the bond load capacity of the adhesive joint increases from 5.2 to 10.5 kN, as bond length increases from 30 to 100 mm; hence the increased 70 mm bond length carries 5.3 kN load capacity. During this stage the addition of bond length will carry load linearly and reduces the stress concentration at the edge. From 100 mm to 130 mm, it has been observed that the 40 mm bond length contributes to increase 0.4 kN load capacity. During this stage, the load induces the stresses at the edges, which are greater than the elastic limit of the adhesive, so elastic plastic transition takes place, and further addition of bond length is not going to have any significance. 100 mm bond length has been considered as the adequate bond length for this experimental condition.

3.3 Vacuum pressure effect on bond strength

Compaction pressure was applied on the wet fabric with the help of the vacuum bag which is under vacuum pressure. An attempt has been made to find the effect of vacuum pressure on the adhesive bond strength of the single-strap SS-CFRP joint. Like in hand/wet lay-up, the fibres were impregnated. Initially the mould is under atmospheric pressure. Different vacuum pressures were introduced in the mould cavity. As the vacuum pressure increases, the compaction pressure acting on wet fibres increases. Hence more amount of adhesive comes out of the patch and is absorbed by the breather fabric. The air voids present in between the layers would also come out along with the excess adhesive. During this study, the other parameters considered are 20% NSA + 80% SA mixture heated to 55°C, 100 mm bond length and sandblasted surface. The prepared samples were tested on UTM for its tensile shear strength.

From **Figure 7**, it was evident that the increased vacuum pressure increases the load capacity of the prepared joint. With 100 mm of Hg vacuum pressure, the obtained joint thickness and load capacity were 4.84 mm and 9.4 kN. With 700 mm of Hg vacuum pressure, the obtained joint thickness and the load capacity were 4.32 mm and 12.6 kN. With more compaction, more amount of adhesive can be

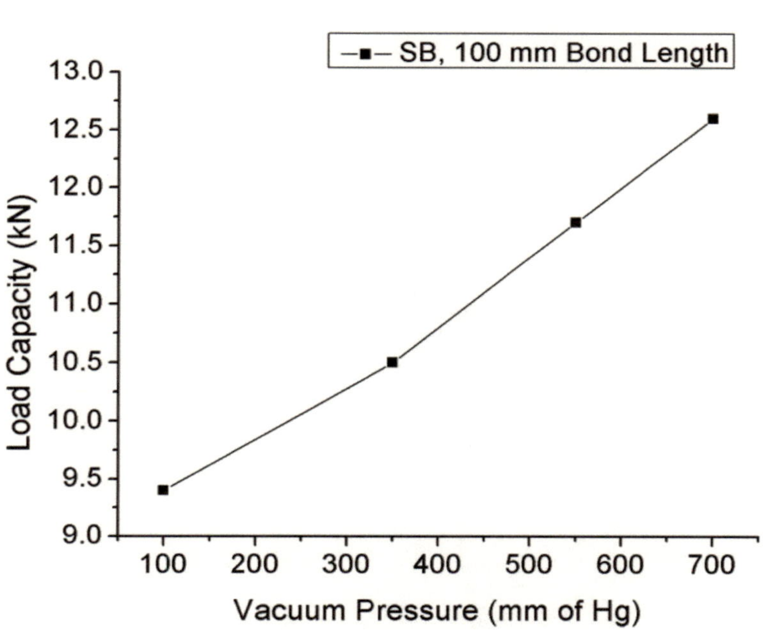

Figure 7.
Load capacity for various vacuum pressures.

squeezed out of the patch which increases the fibre volume fraction and reduces adhesive layer between SS plate and carbon fibre. Hence the adhesive bond strength increases [20].

3.4 Effect of surface texture

Surface preparation is very essential in adhesive bonding; a proper surface may provide a high bond strength. In order to find the proper surface to the adhesive bonding, the adherent surfaces were prepared with plane surface cleaned with acetone, chemical etched surface, sandblasted surface and surface texture created in the form of circular cavities at different densities with a depth of 80 ±5 μm as shown in **Figure 1b**. The surface roughness was measured using a 3D microscope, and the values can be seen in **Table 2**. The surface morphology of the samples was studied using SEM images as shown in **Figure 2**.

The samples for tensile test was prepared with the optimal conditions like 100 mm bond length, 700 mm of Hg and 20% NSA + 80% SA mixture heated to 55°C. Tested results can be seen in **Table 2**. The specimens were failed by delamination between steel and carbon fibre interface (no residues of fibres). The failed surfaces can be seen in **Figure 8**.

From the results, it was evident that the circular surface cavities spread over the 33% of the bonded area were shown a maximum bond strength of 14.15 kN, which is 26% higher than plane surface, 38% higher than etched surface and 12% higher than sandblasted surface. Surface texturing increases the surface roughness of the adherent which in turn increases the mechanical interlocking, and hence the adhesive bond strength increases [11].

S. no	Surface morphology	Area fraction (%)	Ra (μm)	Load capacity (kN)
1	Circular cavities	10	8.81	13.35
		25	20.73	13.83
		33	27.08	14.15
2	Sand blasted	100	0.87	12.62
3	Plane sample	100	0.68	11.24
4	Etched surface	100	0.32	10.25

Table 2.
Load capacity of different pre-bond surfaces.

Figure 8.
Surface of the adherent after failure in tension test.

4. Rehabilitation of pipe

The damaged SS pipes were considered for rehabilitation. The rehabilitation capacity was evaluated using hydrostatic pressure test. Here, a man-made through-all defect with 10 mm diameter was machined over an 200 mm diameter pipe. This through-all hole defect was covered with a two-component solid state adhesive (M-Seal). The pipe was tested for hydrostatic pressure with M-Seal adhesive after 24 hrs. A pressure of 500 ± 30 kPa was observed with M-Seal adhesive.

The pipe surface around the defect was prepared as plane surface cleaned with acetone, etched surface, sandblasted surface and sandblasted surface with circular cavities spread over 33% bonded area. The rehabilitation was done on defect filled with M-Seal adhesive; then the composite patch proposed during the present study as 100 × 100 mm bond area (the fibres in the adhesive bonding of SS plates are aligned in the loading direction with a length of 100 mm) with $[0/90]_3$ carbon fibre layers, 700 mm of Hg and 20% NSA + 80% SA mixture heated to 55°C was applied.

These hydrostatic tests were conducted on the rehabilitated pipes as shown in **Figure 9**, and the results were given in **Table 3**. From the results, it is evident that the failure pressure of a pipe can be changed with surface texture. A maximum of 3852 ± 50 kPa was achieved with a pipe surface prepared with the combination of sandblasting and circular cavities. It is 62.8% higher than the plane surface.

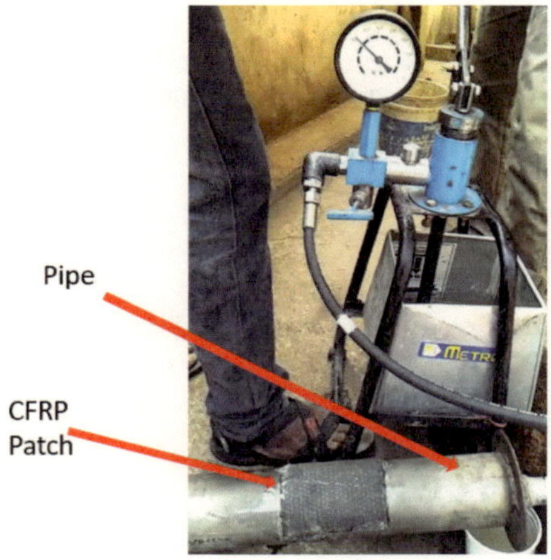

Figure 9.
Pipe hydrostatic pressure testing (left side) and surface prepared before patching (right side).

S. no	Pipe bonded area condition	Failure pressure (kPa)
1	Etched	1525 ± 50
2	Plane	2365 ± 50
3	Sandblasted	3150 ± 50
4	Surface cavities	3852 ± 50

Table 3.
Hydrostatic pressure of pipes.

5. Conclusions

In the present work, the rehabilitation of damaged pipe was done using CFRP patch made with vacuum bagging process. The optimal parameters for rehabilitating the damaged pipe was derived from the experimental results of "the adhesive bonding of SS plates using CFRP patch made with vacuum bagging process". The parameters considered are vacuum pressure, CFRP/adhesive combination and pre-curing temperature, bond length and surface texture. Tensile test was considered to evaluate the bond strength of SS plates, and hydrostatic pressure test was considered to evaluate the bond strength of rehabilitated SS pipes. The experimental data was analysed, and the following conclusions were drawn.

- The viscosity of the adhesive used for the present study was high, and it was reduced with the following methods:
    - The viscosity was changed by adding the low viscous nonstructural adhesive. This addition was reducing the bond strength. Hence it was taken as 20% NSA.
    - Heating of adhesive was shown a significant change in the viscosity and its strength. The strength of the pure adhesive and SS-CFRP joint was increased with increase in curing temperature. But it reduces the gel time. At 55°C the minimum curing time is about 2 hrs which is enough time for the present study.
    - 20% NSA + 80% SA mixture heated to 55°C was considered as adhesive conditions for further study.
- The bond length from 30 to 100 mm might increase the load capacity by approximately 101% (from 5.2 to 10.5 kN). With an increase in bond length from 100 to 130 mm, the load capacity increased by 3.8% (10.5–10.9 kN). Hence, 100 ± 5 could be considered as optimal bond length.
- Vacuum pressure has shown a significant effect on the thickness of samples prepared and the strength of adhesive bonding.
    - At low vacuum pressures, the compaction pressure on the wetted carbon fibre preform is low. Hence a thick layer of adhesive forms between carbon fibre and SS plates which reduces the bond strength.
    - With 100 mm of Hg vacuum pressure, the obtained joint thickness and load capacity were 4.84 mm and 9.4 kN. With 700 mm of Hg vacuum pressure, the obtained joint thickness and the load capacity were 4.32 mm and 12.6 kN.
- The pre-bond surface preparation is a must to have a good adhesive bonding between two adherents. The present study included the surface preparations as sandblasted, chemical etched, plane surface cleaned with acetone and surface texture with circular cavities at different densities.
    - A maximum of 14.15 kN load capacity was observed with the combination of sandblasted surface with circular cavities occupied at 33% of bonded area, whereas the surface with only sandblasting has shown a load capacity of 12.62 kN which is 12% less.
    - The surface roughness of different prepared samples was also studied, and it was evident that the adhesive bond strength is increased with increase in

surface roughness. This is because of the increased mechanical interlocking with roughness of the surface.

- The damaged pipe was prepared with different surface textures and rehabilitated with the optimal parameters obtained with the tensile data of the SS plates. A maximum of 3852 ± 50 was achieved with a pipe surface prepared with the combination of sandblasting and circular cavities. It is 6.7 times higher than M-Seal covered cavity and 62.8% higher than the plane surface.

Vacuum bagging process has many advantages in fabricating composite laminates. The proper use of vacuum bagging process in rehabilitation process may result in strong and eco-friendly adhesive bonded joint.

Acknowledgements

The authors would like to thank "Department of Mechanical Engineering, National Institute of Technology, Warangal", for providing the facilities.

Conflict of interest

The authors have no conflict of interest.

Author details

Syam Kumar Chokka, Beera Satish Ben* and Kowtha Venkata Sai Srinadh
National Institute of Technology, Warangal, India

*Address all correspondence to: drsatishben@gmail.com

References

[1] Lu Y, Li W, Li S, Li X, Zhu T. Study of the tensile properties of CFRP strengthened steel plates. Polymers. 2015;**7**(12):2595-2610

[2] Osouli-Bostanabad K, Tutunchi A, Eskandarzade M. The influence of pre-bond surface treatment over the reliability of steel epoxy/glass composites bonded joints. International Journal of Adhesion and Adhesives. 2017;**75**:145-154

[3] Hollaway LC. The evolution of and the way forward for advanced polymer composites in the civil infrastructure. Construction and Building Materials. 2003;**17**(6-7):365-378

[4] Hollaway LC. A review of the present and future utilization of FRP composites in the civil infrastructure with reference to their important in-service properties. Construction and Building Materials. 2010;**24**(12):2419-2445

[5] White DM, Daniell WE, Maxwell JK, Townes BD. Psychosis following styrene exposure: A case report of neuropsychological sequelae. Journal of Clinical and Experimental Neuropsychology. 1990;**12**(5):798-806

[6] Castillo L, Baldwin M, Sassine MP, Mergler D. Cumulative exposure to styrene and visual functions. American Journal of Industrial Medicine. 2001;**39**(4):351-360

[7] Jeong H. Effect of void on the mechanical strength and ultrasonic attenuation of laminated composites. Journal of Composite Materials. 1997;**31**(3):276-292

[8] Bowles KJ, Frimpong S. Void effects on the interlaminar shear strength of unidirectional graphite-fiber-reinforced composites. Journal of Composite Materials. 1992;**26**(10):1487-1509

[9] Robson JE, Matthews FL, Kinloch AJ. The strength of composite repair patches: A laminate analysis approach. Journal of Reinforced Plastics and Composites. 1992;**11**(7):729-742

[10] Strong AB. Fundamentals of Composites Manufacturing: Materials, Methods, and Applications. 2nd ed. Dearborn: Society of Manufacturing Engineers; 2008

[11] Guo L, Tian J, Wu J, Li B, Zhu Y, Xu C, et al. Effect of surface texturing on the bonding strength of titanium–porcelain. Materials Letters. 2014;**131**:321-323

[12] Hadigheh SA, Kashi S. Effectiveness of vacuum consolidation in bonding fibre reinforced polymer (FRP) composites onto concrete surfaces. Construction and Building Materials. 2018;**187**:854-864

[13] EsnaolaA,TenaI,Saenz-DominguezI, Aurrekoetxea J, Gallego I, Ulacia I. Effect of the manufacturing process on the energy absorption capability of GFRP crush structures. Composite Structures. 2018;**187**:316-324

[14] Zhang D, Heider D, Gillespie JW Jr. Void reduction of high-performance thermoplastic composites via oven vacuum bag processing. Journal of Composite Materials. 2017;**51**(30):4219-4230

[15] Teng JG, Yu T, Fernando D. Strengthening of steel structures with fiber-reinforced polymer composites. Journal of Constructional Steel Research. 2012;**78**:131-143

[16] Hart-Smith LJ. Advances in the analysis and design of adhesive-bonded joints in composite aerospace structures. 1974

[17] Schneider LFJ, Cavalcante LM, Silikas N. Shrinkage stresses generated during resin-composite applications: A review. Journal of Dental Biomechanics. 2010;**2010**:131630

[18] Singh JIP, Singh S, Dhawan V. Effect of curing temperature on mechanical properties of natural fiber reinforced polymer composites. Journal of Natural Fibers. 2018;**15**(5):687-696

[19] Schnerch D, Dawood M, Rizkalla S, Sumner E, Stanford K. Bond behavior of CFRP strengthened steel structures. Advances in Structural Engineering. 2006;**9**(6):805-817

[20] Amirkhosravi M, Pishvar M, Altan MC. Improving laminate quality in wet lay-up/vacuum bag processes by magnet assisted composite manufacturing (MACM). Composites Part A: Applied Science and Manufacturing. 2017;**98**:227-237

Section 2

Applications of Adhesives and Adhesive Joints

Chapter 5

Functional Adhesive Trend for Assembly Industry

ChulSoo Woo

Abstract

Recent applications in the industry require more and more cost-saving, more effective, and more reliable assembly of various substrate parts that are used for end-use product for manufacturers. This chapter can provide an insight on the solutions of different functional bonding and sealing technologies available to manufacturers who have used conventional methods for a long time which help them open their eyes to new solutions that can provide a faster, lighter, lower cost yet achieve more reliable assemblies resulting in more competitive assemblies in the market place. Various technologies provide an optimum solution to different industries and market segments for OEM industry such as appliance, HVAC, filter, medical, construction/agricultural equipment, recreational vehicles, railway, marine, electric motor, loudspeaker, elevator, small engine, valves, pumps, hydraulic system, etc. and for repair industry such as power plant, steel mill, mining, and car repair markets, too.

Keywords: anaerobic, cyanoacrylate, light cure (ultraviolet and visible), acrylic structural (two-part and no-mix), acrylic structural (two-part and pre- and post-mix), silicone, polyurethane

1. Introduction

Adhesives have been a long history ever since the presence of mankind on this earth. It was found out that the jewelry box unearthed from the tomb of ancient Egyptian Pharaoh Tutankhamen had used a glue to assemble it with a beautiful appearance. When the Genghis Khan of Mongolian Empire had conquered the most of Asia and some Europe, the most powerful weapon they had used a bow made of a buffalo bone with laminations bonded with a glue which had become much more powerful than that of enemies. Globally famous musical instrument Stradivarius violin had been found out that a flue was used for laminations that help generate a beautiful sound for generations to come. Through Industrial Revolution, the key trend was how many and how fast they can manufacture within a short amount of time. Nowadays assembly industries have been looking for ways to help how they can manufacture faster yet increase the reliability of their products more. Those functional reactive adhesive and sealants featuring their benefits and typical applications will be introduced in this article in comparison with conventional assembly methods.

They are mainly machinery adhesives (anaerobic), light cure adhesives (LCA), cyanoacrylate adhesives (CA), acrylic structural adhesives, silicone and polyurethane, etc.

2. Overview of adhesive solution

Bonding solution has provided a faster assembly, lighter assembly, and more reliable assembly than various conventional mechanical assemblies. The following section will help you understand the features and benefits of reactive adhesive and sealant technology with the target of explaining the adhesive engineering terminology in order to understand a common basis for the following chapters.

2.1 Universal truth on bonding

Wood is not a proper substrate for welding as it will burn, not melt, plastic is not a proper substrate for soldering or brazing as it will melt away, and steel is not a proper substrate for nail as it will not bond and cause a localized high stress peak. However, adhesive can bond wood, plastic, and steel without burning, without melting and without having a high localized stress peak.

2.2 The adhesive joint

Adhesives are connecting between two substrate surfaces, where they are of the same or different materials. The bonding mechanism depends on the adhesion and cohesion.

2.3 Adhesion

"Adhesion" is an interaction between adhesive and substrate surfaces. There are two main parameters for achieving a good bonding. One is molecular interactions called "van der Waals" force, and another one is a mechanical interaction called "mechanical keyway" in the assembly. Adhesive was applied on the surface flows, due to its rheology, into the valley of surface roughness to fill and grip the surface. This way the mechanical keyway assembly can exhibit a high resistance against external vibration or impact. In case of low surface energy substrates such as Teflon, polypropylene, and polyethylene, adhesive wetting the surface flow into the valley is not easy resulting in improper mechanical keyway effect. In order to overcome this, a surface treatment is required to increase the surface roughness and mechanical keyway effect such as primer, plasma and flame treatment, etc.

2.4 Cohesion

"Cohesion" is an interaction between the molecules of adhesive monomer. This is a combination of "van der Waals" force and molecular reaction of adhesive monomers. This helps to have a toughness of the adhesive. This property is one of the important parameters especially when a reliable structural bonding against impact and vibration is required and multifunctionality of oligomer is used together with elastomers.

3. Various functional adhesives

Most functional adhesives are reactive polymers; in most cases they change from liquid to solid through various chemical polymerization reactions. Numerous functional adhesives have been developed with special curing properties for specific application situations. Those adhesives can be classified into the following groups according to the curing property. They are anaerobic reaction, exposure to light (ultraviolet and visible—also secondary curing option), anionic reaction (cyanoacrylate), activation system (modified acrylics), and moisture curing (silicones, polyurethane).

3.1 Adhesives cured by anaerobic cure

When adhesive is applied onto one side of a metal substrate, adhesives contact with metal ion (Fe^{2+} and Cu^{2+}) for some anaerobic initiator to be broken into free radicals which are unstable, but the presence of air to contact adhesive is stronger; therefore it remains as a liquid. As a next step, when the other mating part comes in contact with the first part with adhesive in between, more anaerobic initiators contact more metal ions to be broken into more free radicals, and at the same time, oxygen contact is blocked by both substrates; then those unstable free radicals react with a monomer nearby which becomes unstable, too, and then it reacts with other monomers nearby, and gradually all the monomers react and become polymerized (cure) (see **Figure 1**) [1].

In case of inactive metal substrates that contain low metal ions such as Cu^{2+} or Fe^{2+} and have a slow reaction with photo initiator and slow cure, an acceleration method such as primer and heating can be used to speed up the curing. Anaerobic adhesives become a very reliable thermoset plastic after cure [2].

3.1.1 Anaerobic adhesive features

This cured thermoset plastic property provides an excellent resistance to various chemical environment, excellent resistance against external impact, and vibrational force due to its mechanical "keyway" effect and high temperature environment up to 180°C and some special formulation is up to 350°C. It also has a controlled viscosity for optimum dispensing and controlled cure speed depending on the requirement of the application. This allows a normal machining by saving a machining cost, and it also allows an automatic dispensing of adhesive for more effective manufacturing.

Curing, especially the reaction speed of anaerobic products, is mainly influenced by the following: substrates, as some are active substrates that contain a rich metal ion, whereas inactive substrates contain less metal ion; the bond line gap, as the reaction takes a short time with a thin bond line, while a thick bond line takes a long time to polymerize; temperature, as elevated temperature activates the reaction of adhesive, while low temperature slows down the reaction of adhesive; and activator, as it will activate the inactive surface as if an active substrate is used.

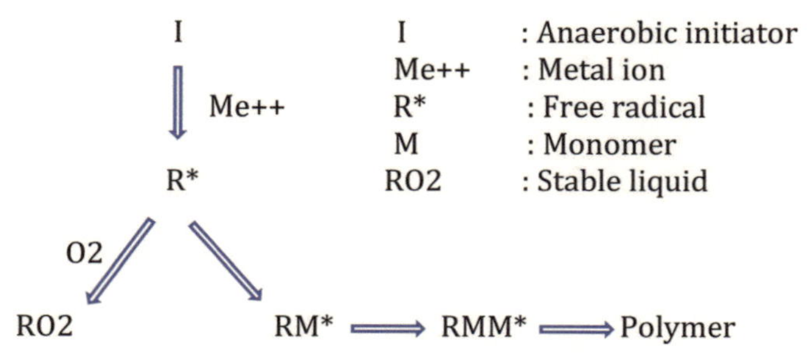

Figure 1.
Anaerobic cure mechanism.

3.1.2 Active and inactive metal substrates

Anaerobic products react fast to active metal substrates and react slow to inactive substrates. Therefore, it is important to understand what type of substrates they have for bonding and sealing application. Active substrates contain more Cu^{2+} and Fe^{2+} ions, while inactive substrates have less Cu^{2+} and Fe^{2+} ions (see **Table 1**) [1].

3.1.3 Anaerobic shelf life

Anaerobic liquid products need a constant oxygen contact to stay as a liquid and to become stable. When a new bottle was opened and the adhesive product level is only half of the container, it is important to know that it is to ensure that adhesive is in constant contact with oxygen in the container. Therefore, if the product's net amount is "250 ml," then the total volume of packing is "500 ml"; likewise in case of a "50 ml" product, the total volume of packaging is "100 ml."

Anaerobic adhesives have a different shelf life depending on the size of the packaging. The smaller the packaging, the longer the shelf life; in other words, the bigger the packaging, the shorter the shelf life. The whole mechanism behind this is that for smaller packaging, the distance oxygen penetrated into the center of adhesive product is shorter, so the whole amount is under the influence of oxygen; however for bigger packaging, the distance of oxygen to penetrate into the center of adhesive product is longer, so the monomers in the center of adhesive products have less amount of oxygen contact and become unstable resulting in increased viscosity and premature reaction of monomers inside the packaging.

3.1.4 Recent technological trend

Recent technological trend of anaerobic adhesive and sealants is to bring the performance of core products to a next level so that those few core products can cover various application requirements; in the past, there were separate products to meet fast cure, high temp resistance, oily surface cutting, active reaction to inactive substrates, etc. However currently, those few core products can meet most of the above requirements at the same time such as this adhesive has a high temperature resistance up to 180°C (constant condition) so it can be used for those applications that could not be used due to high temp requirement; this adhesive has fast reaction to inactive substrates such as stainless steel, aluminum and zinc plated steel, and this adhesives cuts through a slight oil coating on the substrate and it adheres well to the substrate; therefore no special cleaning is required.

Active materials (speed cure)	Passive materials (slow cure)
Brass	Aluminum
Bronze	Ceramics
Copper	Chromate films
Iron	Glass
Steel	High-alloy steel
	Nickel
	Oxide films
	Plastics
	Stainless steel
	Zinc

Table 1.
Active and inactive substrates.

3.1.5 Typical applications

There are four major applications, namely, thread locking application for locking bolt and nut assembly, or stud and tap assembly, in place against loosening; thread sealing application for sealing various fittings and hydraulic and pneumatic thread connections against fluid leakage; retaining application for shaft and gear, bearing, and housing against slippage or free spinning; and gasketing application for machined metal flanges against fluid leakage.

3.1.5.1 Thread locking

3.1.5.1.1 Conventional fastening methods

Those conventional methods used in the assembly industry are flat washer, spring washer, nylon lined nut, serrated bolt, etc.

They have shortcomings such as the following: they loosen under vibration, thermal expansion, and/or plastic deformation, they do not seal thread as the space between bolt and nut is all empty, they require extensive inventory of several fastener shapes and sizes, they are prone to rust and damaged surface due to empty space and poor handling, they have a wide bolt tension scatter as each bolt assembled with the same tightening torque shows a different bolt tension on a flange, they show lower break loose torque than tightening torque resulting in unstable assembly, and they are more costly than liquid thread locking adhesive [3].

3.1.5.1.2 Liquid thread locking adhesive

Those liquid thread locking adhesives have the following benefits: they lock bolt and nut in place, they seal against leakage and internal pressure, they prevent thread corrosion, they provide a controlled lubricity for consistent bolt tension on a flange, they provide a torque augmentation with higher break loose torque than tightening torque, they provide a controlled strength for easy disassembly for repair, they have a low inventory with less storage space, and they have a cost saving [4].

3.1.5.2 Thread sealing

3.1.5.2.1 Conventional thread sealing method

Those conventional methods used in the assembly industry are polytetrafluoroethylene (PTFE) tape, O-ring, pipe dope, etc.

They have shortcomings such as PTFE tape shards generated during pipe assembly contaminate internal hydraulic pipe system, and PTFE tape can have a leak once readjusted after pipe location setting and requires a manual wrapping of time and cost. Another conventional method O-ring has no sealing effect at dynamic environment and requires an additional machining or molding. Pipe dope has a low pressure sealing [5].

3.1.5.2.2 Liquid thread sealing adhesive

Those liquid thread sealing adhesives have the following benefits: they provide a good sealing, they prevent a self-loosening, they cause no contamination inside hydraulic piping system, they cure slowly to allow pipe location adjustment after initial tightening yet provide an instant low pressure sealing, they protect pipe threads against corrosion and galling, and they show a high pressure resistance after cure [6].

3.1.5.3 Retaining

3.1.5.3.1 Conventional retaining methods

Those conventional methods used in the assembly industry are positive drive (pins, keys and splines), friction drive (press fit, shrink fit), and welding/soldering.

They have shortcomings such as the following: they have a notorious back lash that causes an extremely high stress, deformation, and consequent failure of assembly; they have a fretting corrosion that causes a lower torque transmitting capability, and their machining is costly and lengthy to meet the right assembly tolerance; they need additional equipment required for press fit and shrink fit; and in overall they are more complicated, time-consuming, and less reliable [7].

3.1.5.3.2 Liquid retaining compound adhesive

Those liquid retaining compound adhesives have following benefits such as they are less costly in machining as they don't need a tight tolerance; they cause no back lash, and therefore it allows a high durability and reliability of the assembly; they allow the assembly assembled simpler and easier; and they can be used to conventional designs if they can't change the design and it can increase the durability of the assembly, too [8].

3.1.5.4 Gasketing

3.1.5.4.1 Conventional gasketing methods

Those conventional methods used in the assembly industry are various compression gaskets (pre-cut gaskets, O-ring seals and profile packings).

They have shortcomings such as the following: they achieve a sealing by compression for a long time which is not realistic; they have a shrinkage, creep, and compression set under high flange pressure; they require to store multiple shapes to seal different flange shapes; they are costly with an intricate flange shape; they are susceptible to external or internal impact or pressure; and they require additional machining or smooth machining to achieve a sealing with a cut gasket [9].

3.1.5.4.2 Liquid gasketing sealant

Those liquid gasketing sealants have the following benefits: they have a metal-to-metal contact, thus becoming a very stable assembly, they can be used to all different shapes of flange with one product, they can provide a unitized assembly with "keyway effect" between flanges, they can eliminate all the mistakes that occur with the manual operation, they can be used for automatic dispensing and assembly, and in overall they can increase the durability, reliability, as well as the cost reduction of the assembly [10].

3.2 Adhesives cured by light (ultraviolet and visible)

The cure time of these adhesives depends on the intensity and on the wavelength of the light. Polymerization initiated by the light thus always requires a proper curing property of the adhesive and correct exposure to a light radiation. When a light cure adhesive is applied and goes through the area of light irradiation, the photo initiators in the adhesive are split to become free radicals which are very unstable. In this situation, unlike to anaerobic cure, there is no condition that

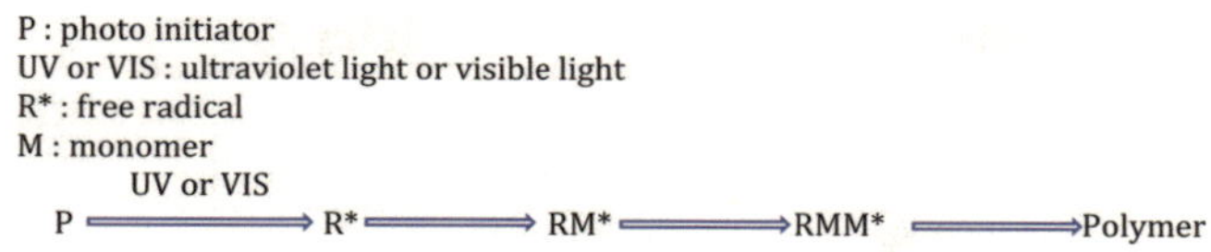

Figure 2.
Light curing mechanism.

prevents a curing of the adhesive as the power of light energy is so high to overcome all the other hindrance. The free radicals formed in turn start the polymerization by reacting with a monomer nearby, and it becomes unstable, too, and then reacts with the next monomer and finally polymerized in a very fast speed (see **Figure 2**). It is critically important to use a right light curing lamp equipment that emits a proper wavelength of radiation spectrum ideally matched to the curing properties of light cure adhesives [11].

There are three types of light cure adhesives, namely, UV cure adhesive, UV+ visible cure adhesive, and visible cure adhesive.

3.2.1 Ultraviolet (UV) light cure adhesive

UV light is in the range of 40–400 nm wavelength. The whole UV wavelength does not necessarily cure adhesive. A specific wavelength of UV light affects the adhesive. UVC is a short wavelength, and it cures the surface of adhesive as it is a short wavelength and not easy to penetrate the adhesive inside; therefore it is used for surface cure of adhesive. UVA is a long wavelength, and it cures the volume of adhesive as it is a long wavelength and easy to penetrate the adhesive depth; therefore it is used for a depth cure. Care should be taken when selecting a clear plastic for light cure. Some plastics such as UV stabilizer filled PC looks like a clear plastic to the eyes of people, but UV light cannot go through the plastic as the UV stabilizer inside the PC can absorb the UV light and cannot reach the adhesive bond line. It is also critical to know that the temperature inside UV curing chamber or conveyor should be kept less than 60–70°C depending on the thermal sensitivity of parts exposed to UV light as the UV light bulb can irradiate not only UV light but also emit infrared (IR) and heat. Otherwise temperature can reach above the limit of plastic parts. UV adhesive was the first-generation light cure adhesive introduced in the assembly industry.

3.2.2 Ultraviolet (UV) light + visible light cure adhesive

Ultraviolet (UV) + visible light is in the range of 40–405 nm wavelength. UV photo initiator had a limited cure through volume (CTV) capability, and some visible light initiators were added together to achieve better CTV and faster curing with less UV dependence that was harmful to operators. The photo initiator for visible light at the time had a limited (partial) visible light wavelength coverage. This is a second-generation light cure adhesive introduced in the assembly industry.

3.2.3 Visible light cure and adhesive

Visible light cure is in the range of 400–1000 nm wavelength. This is only for visible light cure adhesive, and it can cover wider visible light wavelength coverage; therefore it can cure through different colored clear plastics which previous light cure adhesive was not able to cure as it cannot go through the colored plastics. It also can cure CTV 12 mm deep as it has a longer wavelength. The visible light cure

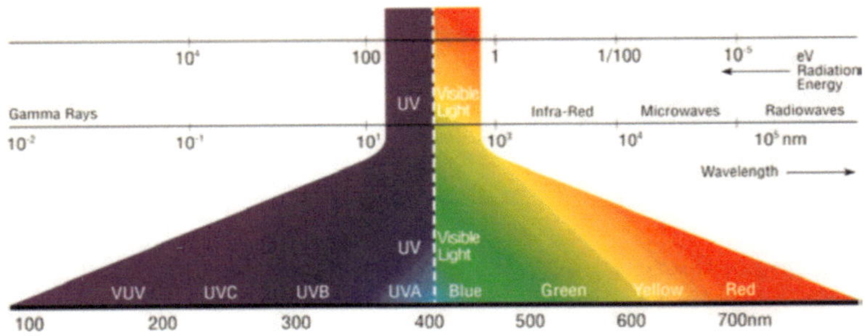

Figure 3.
Light wavelength.

lamp is safer than UV lamp as it emits no UV light. Therefore, there is also no need to wear safety UV block glasses and skin protection wear as well as it has lower cost than UV lamp (see **Figure 3**) [11].

3.2.4 Light curing lamp equipment by the application type

Light curing lamps by the application type consist of benchtop lamp that can be used for relatively big irradiation area of small batch production or testing at development lab, conveyor type that can be used for relatively big irradiation area of volume production, and fiber optic/pencil type that can be used for specific small irradiation area of development lab and small and volume productions that require a localized area curing.

3.2.5 Light curing lamp equipment by the lamp type

Types of light curing lamp consist of medium pressure mercury arc lamp, high pressure mercury arc lamp, metal halide lamp, fusion lamp, and LED lamp.

Also, light curing process demands are divided into three types, namely, depth curing by light radiation, surface curing by light radiation, and secondary cure systems.

3.2.6 Depth curing

To cure adhesive to maximum depth, light curing systems which emit high intensity light in the wavelength in the band from 300 to 400 nm (longer UV wavelength, UVA light) should be specified. This is useful for applications of potting or bigger gap and big fillet cure bonding [11].

3.2.7 Surface curing

Surface curing is especially important when potting or bonding with light cure adhesives are done. If adhesive surface becomes tacky after light cure, it can contaminate the adjacent area and sensitive components by collecting dust and outgassing of uncured adhesive. Wavelength at 254 nm is the one that can achieve a proper surface cure. A selection of proper light cure lamps that emit a proper wavelength is crucial for successful curing application. It is worthwhile to reiterate that this is very important for the application of surface coating or fillet cure to prevent contamination from tackiness on the cured adhesive surface (see **Figure 4**) [11].

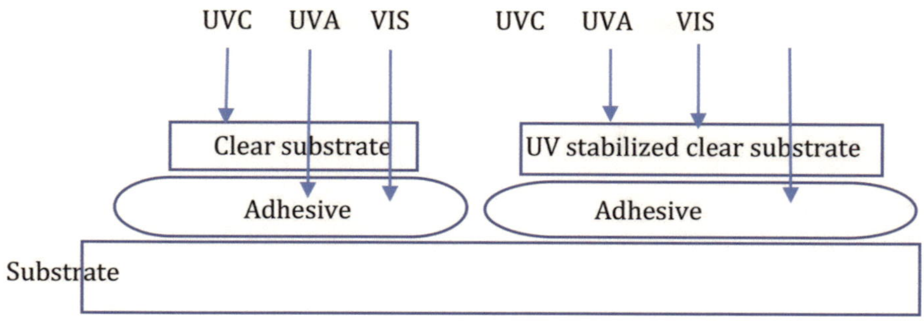

Figure 4.
Light penetration to different substrate.

Also as shown in **Figure 4**, care should be taken to select a right light curing lamp to some special clear plastics. One may think that UV stabilized clear plastics can be cured by UV light curing lamp because it looks clear to our eyes.

But our eyes use a visible light to see its clearness. Such clear plastics contain UV stabilizer that blocks the UV light that cannot reach out to the adhesive. It is all absorbed by the UV stabilizer in the plastic. In that case, visible light cure adhesive can be used to cure adhesive through the UV stabilized clear plastics [11].

3.2.8 Curing by secondary mechanisms

In numerous applications the light does not reach all areas of adhesive bond line due to opaque substrates. Therefore, other functional light cure adhesives have been developed with a secondary cure system for areas not exposed to the light such as anaerobic curing, heat, ambient moisture, and activators.

Also there are various dual cure adhesive technologies, namely, light cure anaerobic, light cure CA, light cure acrylic, light cure silicone, and light cure epoxy.

3.2.9 Light cure adhesive characteristics

Light cure adhesives have the following characteristics: on demand cure, high strength, high gap filling capability, very short curing time in handling strength, good to very good environmental resistance, and good dispensing capacity with automatic application systems as a single-component adhesive.

3.2.10 Typical applications

Typical applications of light cure adhesives such as bonding glass to itself or to metal, bonding transparent plastics, sealing electrical components (e.g., relays), bonding electrical components to printed circuit boards (PCBs), conformal coating of PCBs, bonding/sealing in high temperature applications, retaining metal parts + cure excess adhesive by light for fast fixturing, bonding metal and plastic parts + cure excess adhesive by light for fast fixturing, and magnet bonding with a fillet cure.

3.3 Adhesives cured by anionic reaction (cyanoacrylates)

Single-component cyanoacrylate adhesives polymerize on contact with slightly weak base (alkaline) surfaces. In general, ambient humidity in the air and on the substrate surface is sufficient to initiate curing within a few seconds. This is called "anionic" polymerization (see **Figure 5**) [12].

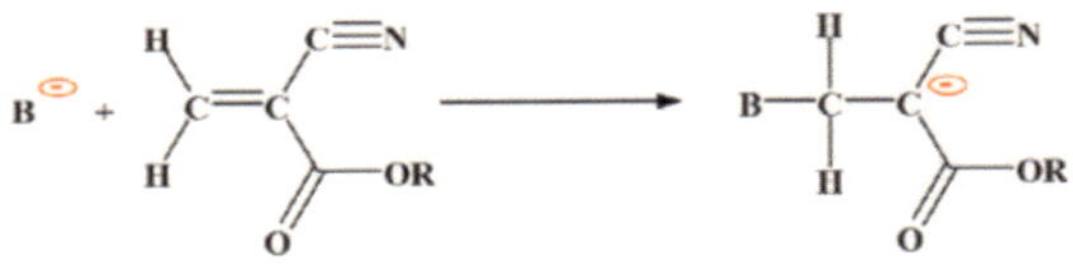

Figure 5.
CA anionic reaction 1.

B-C-C + C=C → B-C-C-C-C

Figure 6.
CA anionic reaction 2.

The adhesive contains acidic stabilizer, and when adhesive was applied to a substrate surface, the acidic stabilizer that keeps the adhesive stable reacts with moisture on the surface. As the acidic stabilizer is consumed by moisture on the surface, the adhesive becomes unstable, and monomer reacts to each other, and the negative partial load is then passed to the next monomer in a chain reaction as seen in **Figure 6** [12].

The polymerization continues until no monomer is available or a molecule with positive partial load (e.g., stabilizer) is reached.

3.3.1 General characteristics

3.3.1.1 Rule of thumb

The rule of thumb for CA adhesives such as the lower the viscosity, the faster the curing speed; the higher the viscosity, the slower the curing speed, and it generates while curing(exothermic reaction); the thinner the bond line, the faster the curing time, and the bigger the package size, the longer the shelf life, and not recommended for glass or glazed ceramics, and not ideal for a large area bonding, and liquid monomer can cause a stress crack on some sensitive plastics, and some product may cause a blooming.

3.3.1.2 Cure versus bond line

To achieve the fastest cure to handling strength, a "zero gap" condition is desirable.

3.3.1.3 Cure versus moisture

The best results are achieved when the relative humidity in the working environment is 40–60% at room temperature. Lower humidity leads to slower curing; higher humidity accelerates it but may impair the final strength of the bond. Longer cure times, however, slow down production. With the help of a humidity control system, favorable humidity levels can be kept constant in the bonding workplace.

3.3.1.4 Cure versu surface pH

Acidic surfaces (pH value <7) may delay or even prevent curing, whereas alkaline surfaces (pH value >7) accelerate curing.

3.3.1.5 Cure versus open time

Even though cyanoacrylate fixtures and cures are relatively fast, there are various types with different fixture times depending on application requirements because some applications require a very fast fixture and others require some time to adjust the parts location then fix it. It is not recommended to apply too much adhesive as it needs more time to cure and will slow down the fixture time.

3.3.1.6 Cure versus activator

As cyanoacrylate cures by moisture, the lack of moisture causes a slow cure of adhesive. In this situation, accelerator can be used to speed up the cure even in dry environment. Accelerator can be applied to substrate before adhesive application or can be applied after adhesive application. Both ways can speed up the cure. The latter is often used when some part is secured by adhesive in an open area without two-part bonding such as wire tacking on a PCB board, etc.

When bonding a difficult to bond plastic or rubber, polyolefin primer can be used to enhance the bonding reliability.

3.3.1.7 Storage

The proper storage of cyanoacrylate adhesive is crucial for the best product shelf life. Refrigeration (4C) in an unopened state is recommended. If adhesive bottle opened already, it is recommended to close the bottle tightly and keep it at room temperature and continue to use it. It is not necessary to be refrigerated.

3.3.2 Cyanoacrylate adhesive characteristics

Characteristics of cyanoacrylate are the following: single-component, very fast fixture time (fixturing in seconds) and curing speed, almost all materials may be bonded, wide range of viscosities, very high shear and tensile strength, good aging resistance, and easy automation with dispensing equipment as a single component.

3.3.3 Cyanoacrylate by the monomer chemistry

The monomer chemistry of cyanoacrylate consists of methyl grade, ethyl grade, and alkoxy grade.

3.3.3.1 Methyl grade

This is the first-generation cyanoacrylate. Just like the chemistry, the molecular weight of “methyl” is the smallest, so the number of molecules that can be attached to the bonding surface is the most which means it has a high bonding strength but as the total weight is the lightest therefore it is easy to evaporate and stimulate people’s nose as a strong smell and cause a “blooming” phenomenon, too (see **Figure** 7 and **Table 2**) [12].

3.3.3.2 Ethyl grade

This is the second-generation cyanoacrylate. Just like the chemistry, the molecular weight of “ethyl” is medium, it means the number of molecules that can be attached to the bonding surface is medium, consequently it has a medium bonding strength but as the total molecular weight is medium therefore it is still not light and not easy to evaporate and stimulate people's nose as a less strong smell and cause less "blooming" phenomenon (see **Figure 8** and **Table 2**) [12].

3.3.3.3 Alkoxy grade

This is the third-generation cyanoacrylate. Just like the chemistry, the molecular weight of "methyl" is the biggest, so the number of molecules that can be attached to the bonding surface is the smallest which means it has a low bonding strength, but as the total weight is the highest, therefore it hardly evaporates, has no smell to people's nose, and causes almost no "blooming" phenomenon (see **Figure 9** and **Table 2**) [12].

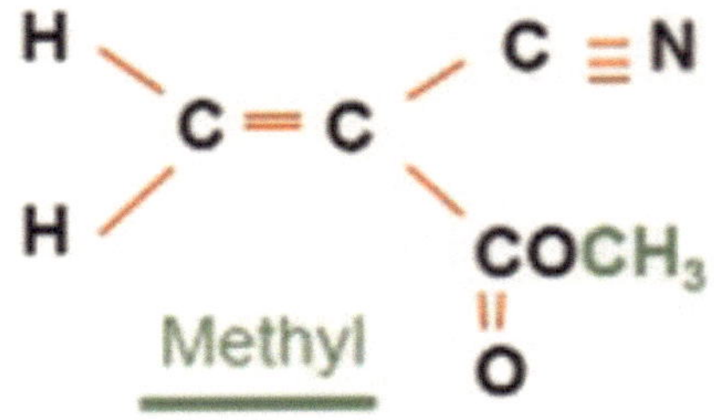

Figure 7.
Methyl grade molecule.

R (alkyl)	Odor	Adhesion	Temp. resistance
Methyl	High	High	High
Ethyl	High	High	High
Alkoxy	Moderate	Moderate	Moderate

Table 2.
Property difference among CA grades.

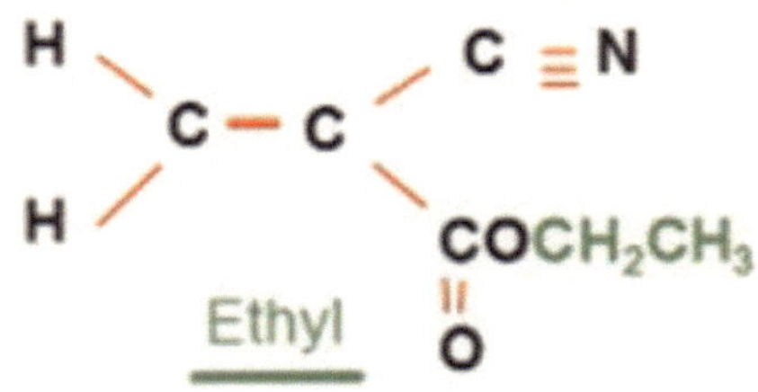

Figure 8.
Ethyl grade molecule.

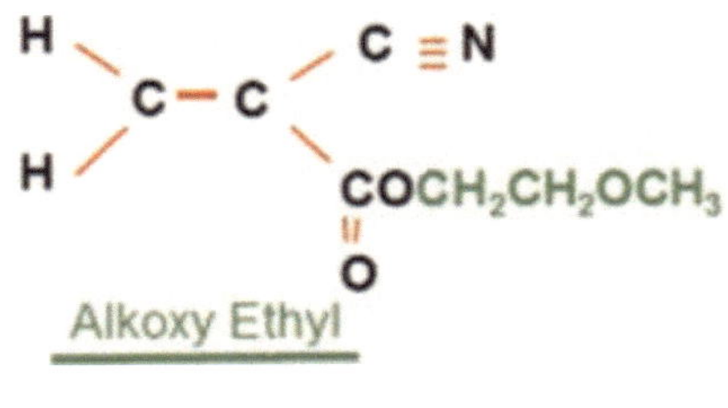

Figure 9.
Alkoxy grade molecule.

3.3.4 Recent trend of CA adhesive properties

Recent bonding performance improvements achieved are for fast fixture time for porous and acid surface, good impact resistance, high temperature resistance, and more clean appearance of bonded assembly.

3.3.4.1 Surface insensitive grade

Typical CA products are surface sensitive which means no effective bonding to porous substrates and acidic surface. It wicks into the porous area and leaves nothing on the surface, so there is no good bonding and the curing time is slow when applied to acidic surface. This surface insensitive grade when applied to porous substrate stays on the porous surface to bond well and, due to some special chemistry, bond fast to the acidic surface. They have a higher temperature resistance and more surface insensitive property (faster fixture time), this is ethyl grade [12].

3.3.4.2 Toughened grade

Typical CA products are not tough which means there is no effective resistance against impact and vibration. It breaks very easily when the assembly is exposed to impact. This toughened grade contains some elastomer inside, so when bonded the bond line absorbs the external impact and maintains a very good bonding. They have a higher temperature and moisture resistance and have surface insensitive property (fast fixture), too. This is ethyl grade [12].

3.3.4.3 Low bloom/low odor grade

Typical CA products have a strong odor and "blooming" which means there are no good working environment and no good assembly appearance quality. This low bloom/low odor grade due to a heavy molecular weight has no pungent smell and almost no blooming. They have a higher temperature resistance and surface insensitive property (fast fixture), too. This is alkoxy grade [12].

3.3.5 Typical applications

Typical applications of cyanoacrylates are used broadly in the various assembly industries and can cover various applications and their requirements such as bonding of porous substrates, acidic surfaces, difficult to bond plastics, dissimilar materials, flexible materials with very fast fixturing, high impact and vibration resistance, high temperature resistance (120C), high humidity resistance, high heat aging resistance, high optical appearance as well as bonding of the most general metal, plastic and elastomer substrates.

3.4 Adhesives cured with activator systems (modified acrylics)

These adhesives are regarded as structural adhesives. Structural adhesive is different from others essentially in the bond of the adhesive, in the monomer type, and in the cure chemistry. The compositional differences result in differences in cure characteristics and in surface affinity and in basic cured properties such as durability of adhesive. Structural adhesives are strong and tough as they possess a high cohesive and adhesive strength. The cohesive strength represents the ability of the material to tolerate stress without failure. Adhesive strength is the ability of the material to stick to the surface. Cohesive failure will leave adhesive on both substrates. Adhesive failure occurs at the interface of the adhesive and the substrate and after debonding, one side substrate will have no cured adhesive left bonded while all the cured adhesive will be left bonded to the other side substrate. These adhesives cure at room temperature when used with activators. Depending on the adhesive, for two-part no-mix type, both the adhesive and activator can be applied separately to the bonding surfaces. It does not react as long as they are not mated allowing some flexibility of application condition. After mating both surfaces with a pressure, reaction starts and adhesive cures to a tough structural adhesive. For two-part premix type, product is dispensed through a double cartridge by a double plunger, and it is mixed and dispensed through static mixer. This way, separate mixing procedure is not required. A precise amount is mixed by volume. For two-part separate dispensing type, adhesive parts A and B are dispensed through a separate nozzle on to the substrate bead by bead or bead beside bead. This way the product is not cured after dispensing, and when the two substrates are mated, the adhesive will be mixed by pressing movement and cures to a tough acrylic adhesive. This is useful when the high-speed production and wide mixing ratio tolerance are required. Three types can be selected properly according to application requirements.

3.4.1 Structural adhesive characteristics

Structural adhesive characteristics are very high shear and tensile strengths, good impact resistance, wide operating temperature range (−55°C to +120°C), almost all materials can be bonded, good gap-filling capacity (especially premixed acrylics), and good environmental resistance.

3.4.2 Structural adhesive (two-component no-mix)

Some grades have special oligomers to provide more toughness. Other grades have multifunctional acrylate and elastomeric oligomer. They are minimally crosslinked. They have a big portion of elastomeric toughening agent, and they generate free radical through redox reaction. The following is the typical oligomeric structure of structural acrylic adhesives (see **Figure 10**) [13].

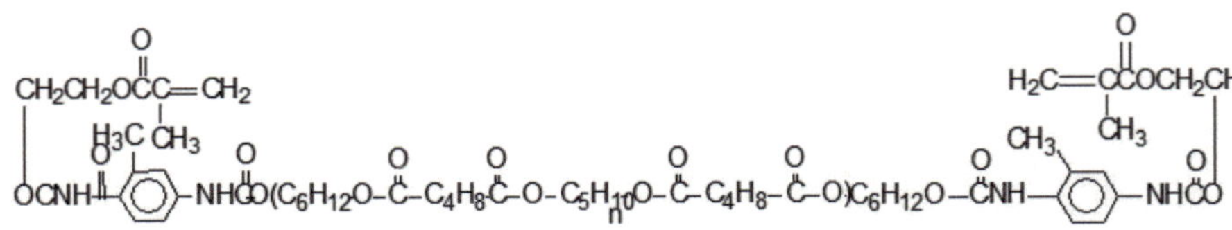

Epoxy Acrylate

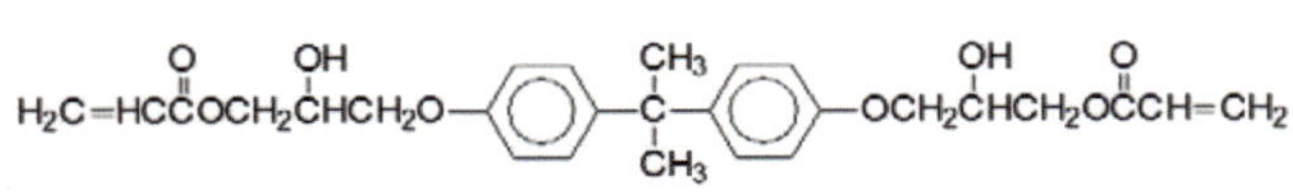

Polyester Acrylate

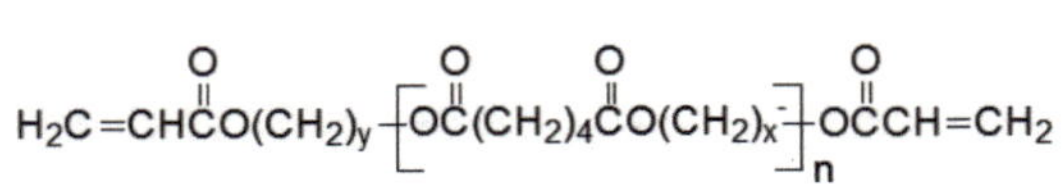

Figure 10.
Different acrylate molecular structures.

Adhesives with multifunctional oligomers cure faster than monofunctional oligomer. Multifunctional oligomers have faster cure speed, better surface cure, better durability, better hardness, and much low odor and toxicity than monofunctional oligomer.

Elastomers are responsible for flexibility and toughness; therefore the structural adhesives result in a high viscosity, and many different types of elastomers are used in this adhesive.

3.4.2.1 Product selection criterion

The following requirements have to be considered when selecting a right adhesive such as fixture time, ultimate bond strength, on-part life, toughness, temperature resistance, and carrier solvent.

3.4.2.2 Typical applications

Typical applications of structural adhesive are magnet bonding for different types of electric motors, magnet bonding for loudspeakers, security alarm sensor bonding, metal structure bonding, flat coil bonding, ferrite core bonding, and voice coil magnet plate bonding for hard disk drive.

3.4.3 Structural adhesive (two-component premix: MMA)

Methyl methacrylate (MMA) adhesives are structural adhesives which are toughened by special elastomers which are "super tough" because they result in a graft polymer that has both excellent peel and impact strength. They cure at room temperature. Their two-component cure is based on redox system. They have a good performance on most substrates especially plastics. They have a good performance on dissimilar substrates. They have a good adhesion to fiber-reinforced plastic (FRP) and gel coat with little or no surface preparation. They have a high modulus and toughness. They have an off-ratio tolerant within 20% during mixing. However, cure rate is temperature dependent, such as in summer and winter, and heat of cure and shrinkage can cause a distortion to read-through at FRP panel.

They will cure when adhesive (part A) and activator (part B) are statically mixed. They will fillet cure and cure speed is dependent on mass. Plastics that are solvated by MMA will fixture fast. The type of mixing ratio is ranging from 1:1 to 10:1. Some products which contain glass or polymer beads that can control bond line thickness are ranging from 0.1 to 0.8 mm diameter.

3.4.3.1 Glass (or polymer) beads

For some special applications, if MMA is used with "0" gap, then the assembly cannot take the advantage of "toughness" of MMA. By using beads, the cohesive strength and impact strength become very high as the bond line absorbs all the external impact. Beads also ensure a controlled exotherm reaction in the bond line, too. **Figure 11** shows a certain bond line gap achieved with spacer. This design

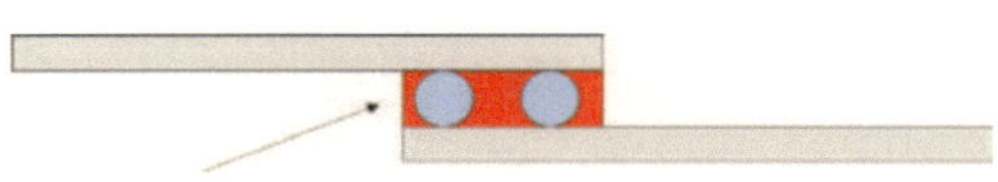

Figure 11.
Structural bonding with spacer.

provides a very high toughness to the assembly as a very high bond line stress is eliminated with this design (see **Figure 11**) [14].

3.4.3.2 Typical applications

Typical applications are galvanized steel bonding for construction to prevent corrosion bonding failure; school bus structure bonding for fire wall, side wall, and roof; marine yacht bonding for deck, stringer, hull, and gel coat; boom bonding for special vehicles; structure bonding with spot welding; office cubicle bonding; composite structure bonding; and case bonding of hand-held display device.

3.5 Adhesives cured by ambient moisture

These adhesives and sealants are polymerized by (in most cases) a condensation reaction which involves reaction with ambient moisture. Two general chemistry types fall into two categories which are silicone and polyurethane.

3.5.1 Silicone

Depending on the curing mechanism, it consists of one-component RTV silicone and two-component silicone. For one-component silicone, it vulcanizes at room temperature either by reaction with moisture in the air and substrates. It starts to cure from the surface and cures into depth. Depending on the type of formulation, it generates a by-product and evaporates. For two-component silicone, it vulcanizes at room temperature by part A and B reactions. Its curing starts from both surface and inside of mixed product. Therefore, its curing is not affected by the lack of moisture in the air.

3.5.1.1 One-component RTV silicone

This is widely used in various industries. Depending on the type of formulation, a by-product is generated such as vinegar smell and non-vinegar smell. Acetic acid type has a pungent odor and is mainly used in the construction area and some machinery repair applications. Oxime type is the most popular in the assembly industry as it has no pungent odor and is regarded as non-acetic acid silicone. The cure speed of silicone primarily depends on the relative humidity which means it is fast in summer time and slow in winter time (see **Figure 12**) [15].

Because of the nature of the curing mechanism, silicones vulcanize from the outside to the inside of the bond line. Due to the necessary moisture migration to

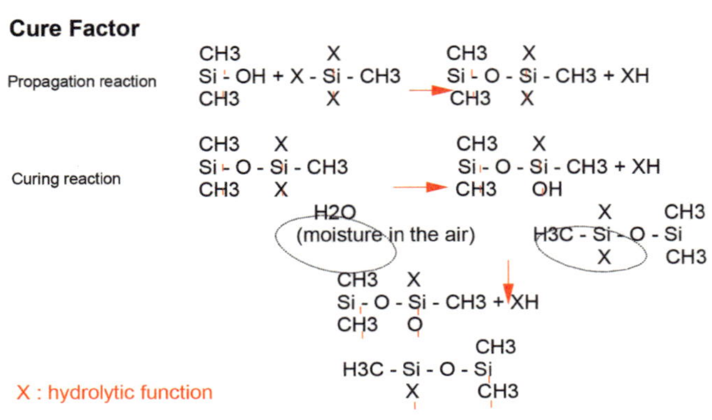

Figure 12.
RTV silicone curing mechanism.

the site of crosslinking, once surface is cured, moisture becomes more difficult to penetrate the already cured layer of silicone to react with internal uncured silicone; therefore the whole curing takes relatively a long time. The depth of cure is limited to 10–15 mm.

3.5.1.1.1 Silicone characteristics

One-component silicone characteristics are the following: excellent thermal resistance ("up to 270°C"), flexible, tough, high elongation, low to medium modulus, effective sealants for a variety of fluid types, and excellent gap filling.

3.5.1.1.2 Typical applications

One-component silicone applications are gasketing and sealing in automotive industry, sealing in high temperature applications, sealing and bonding for small parts, sealing and potting, and coating of PCBs.

3.5.1.2 Two-component silicone

When part A and part B are mixed in a proper mixing ratio, a chemical reaction occurs initiated by the catalyst reacting with water. This chemical reaction occurs not only at the surface but also inside the mixed product. Therefore, it has a very fast fixturing and curing time for faster fixture and assembly, and it allows a structural bonding assembly in more compact assembly line and automated assembly with less manpower. This technology can replace various assemblies that have used double-sided tape or RTV silicone that requires a manual and lengthy assembly line and time with much manpower. During the chemical reaction, moisture ("H_2O") comes in and alcohol ("OH") that was generated during chemical reaction as a by-product comes out and evaporates (see **Figure 13**) [15].

3.5.1.2.1 Silicone characteristics

Two-component silicone characteristics are such as fast tack free time, fast fixture time and fast full curing time. Its curing speed is not affected by moisture and bond line thickness. It is flexible, tough, high modulus and UL 746C listed for high temperature applications. It also has an excellent gap filling capability and high temperature resistance up to 270C. Three colors such as black, gray and almond are available.

3.5.1.2.2 Typical applications

Typical applications of two-component silicone are glass bonding for refrigerator, washing machine, microwave, cook top, dish washer, air conditioner and bridge structure bonding.

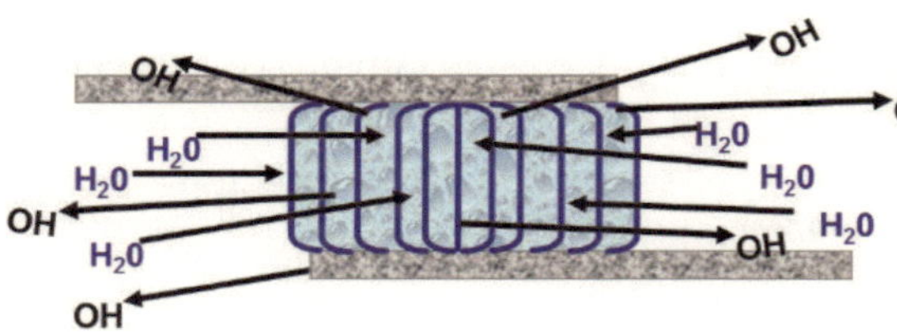

Figure 13.
Two-component silicone curing mechanism.

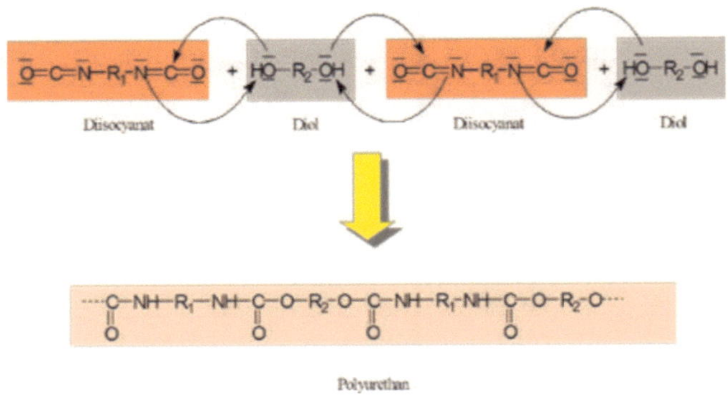

Figure 14.
Curing mechanism of two-component polyurethane.

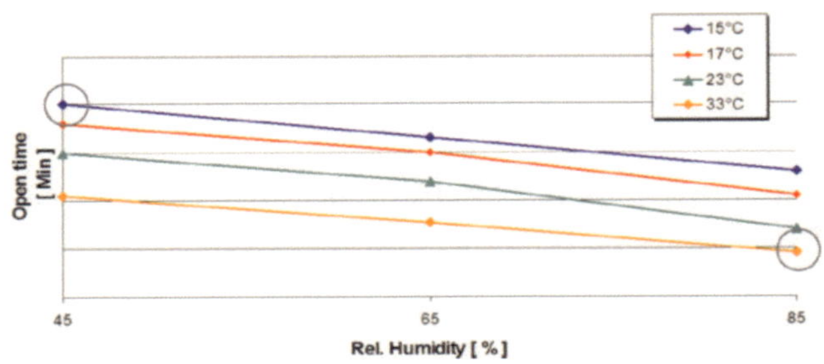

Figure 15.
Open time versus ambient temperature of two-component polyurethane.

3.5.2 Polyurethane

Polyurethane adhesives cure with polyol and isocyanate at room temperature. During the reaction, it generates some carbon dioxide (CO_2). One unique feature of polyurethane adhesive is that its cure is sensitive to the moisture level. For some applications to achieve the best and most durable adhesion, the use of appropriate cleaners and primers (adhesion promoters) is recommended. Depending on the substrates, different primers are used. According to where it is used, they consist of one-component PU and two-component PU. PU has a different open time according to the ambient temperature and mixed amount. **Figures 15** explains that the open time in high temperature environment such as summer time reacts fast and has a short open time while it does a long open time in cold environment such as winter time (see **Figures 14** and **15**) [16].

3.5.2.1 Polyurethane characteristics

Polyurethane characteristics are excellent toughness and flexibility, high elongation, excellent gap filling, paintable when cured, and excellent chemical resistance.

3.5.2.2 Typical applications

Typical applications of polyurethane are bonding/sealing of direct glazing window for automotive, railway, and earth movers (one-component, two-component);

panel lamination bonding for recreational vehicle and mobile home; wind blade bonding (one-component, two-component); sport goods lamination bonding (two-component); and filter bonding or casting for water filter, air filter, mechanical filter, and blood filter (two-component).

4. Conclusion

Major reactive functional adhesives and sealants have been described with conventional methods and benefits of adhesive and sealant solutions in this chapter. The assembly industry still has numerous area that can improve their assembly process with cost saving and increase reliability of their assembly. It would be worthwhile that all the information be thoroughly reviewed and studied, and with this material, it is hoped that ideas to improve a cost and to increase a reliability of various assemblies with such adhesives and sealants would be materialized as a life time guide.

Author details

ChulSoo Woo
Henkel (China) Investment Co., Ltd, Shanghai, China

*Address all correspondence to: cs.woo@henkel.com

References

[1] Woo C, Rudolf N. Anaerobic Technology. Henkel Corporation; 2017

[2] Alan L, Rudolf N. Loctite Worldwide Design Handbook. 2nd ed. Henkel Corporation; 1998

[3] Woo C, Rudolf N, Robert D. Conventional Method of Thread Locking. Henkel Corporation; 2017

[4] Woo C, Rudolf N, Robert D. Loctite Solution for Thread Locking Technology. Henkel Corporation; 2017

[5] Woo C, Rudolf N, Robert D. Conventional Method of Thread Sealing. Henkel Corporation; 2017

[6] Woo C, Rudolf N, Robert D. Loctite Solution for Thread Sealing Technology. Henkel Corporation; 2017

[7] Woo C, Rudolf N, Robert D. Conventional Method of Retaining. Henkel Corporation; 2017

[8] Woo C, Rudolf N, Robert D. Loctite Solution for Retaining Technology. Henkel Corporation; 2017

[9] Woo C, Rudolf N, Robert D. Conventional Method for Gasketing. Henkel Corporation; 2017

[10] Woo C, Rudolf N, Robert D. Loctite Solution for Gasketing Technology. Henkel Corporation; 2017

[11] Woo C, Hermann H. Light Cure Technology. Henkel Corporation; 2012

[12] Woo C, Hermann H. Cyanoacrylate Technology. Henkel Corporation; 2012

[13] Woo C, Ed F. Structural Acrylic Technology. Henkel Corporation; 2008

[14] Woo C, Ed F. Methyl Methacrylate (MMA) Technology. Henkel Corporation; 2008

[15] Woo C, Michael L. Silicone Technology. Henkel Corporation; 2006

[16] Woo C, Michael G. Polyurethane Technology. Henkel Corporation; 2011

Chapter 6

Pressure-Sensitive Adhesive Joints

Jaber Khanjani

Abstract

Self-adhesive materials are called, in the adhesive industry, pressure-sensitive adhesives (PSAs). PSAs are designed in the shapes of latex, solvent borne resin, rubber solution, or hot melt and are being used for foils and films, tapes, labels, and notepads which can make permanent, removable, or semi-removable adhesive joints for applications of general purpose product assembly by simple contact under light pressure. This special class of adhesives does not undergo any physical transformation or chemical reaction during the bonding process. The end-use properties of PSAs result from the nonlinear viscoelastic behavior of the adhesive material, and the elastomeric polymer basis of PSAs imparts them such a viscoelastic behavior caused by a carefully chosen polymer architecture and monomer composition with the proper addition of small molecules called tackifying resins. They are safe to use and easy to handle and thus are increasingly replacing more conventional types of adhesives. In this chapter, we review adhesion mechanism of PSAs, types of PSAs, adhesion properties and tests, mechanical behavior of joints, and especially different aspects of PSA applications.

Keywords: self-adhesives, pressure-sensitive adhesives, viscoelastic behavior, tackifying resins, adhesion mechanism

1. Introduction

Pressure-sensitive adhesives provide a quick and effective fastening method compared to most mechanical joining techniques. Adhesive bonding has become a preferred fastening technology in applications where dissimilar materials need to be joined. PSAs, with a viscoelastic nature, are composed of several key elements: a fluid, adhesive, tackifier and some kind of structured backing (**Figure 1**) [1].

PSAs have generally an elastomer basis, which can be of the following materials such as natural rubber, vinyl ethers, acrylics, butyl rubber, styrene block copolymers, silicones and nitriles. A tackifier, a low molecular weight resin, is added to the elastomer to increase adhesion and is generally one of the following common resins: terpenes, aromatic resins, hydrogenated hydrocarbon resins and terpene-phenol resins [2].

PSA's in the shapes of latex, solvent borne resin, rubber solution or hot melt are used for foils and films, tapes, labels and notepads which can make permanent, removable or semi-removable adhesive joints for applications of general purpose product assembly, display, tags, labels, and mailers, packaging tapes, masking and paper tapes, tapes for hygiene applications, double-sided tapes, cloth and duct tapes, heavy-duty and reinforced tapes and electrical tape as well as a wide variety of other products (**Figure 2**) [3].

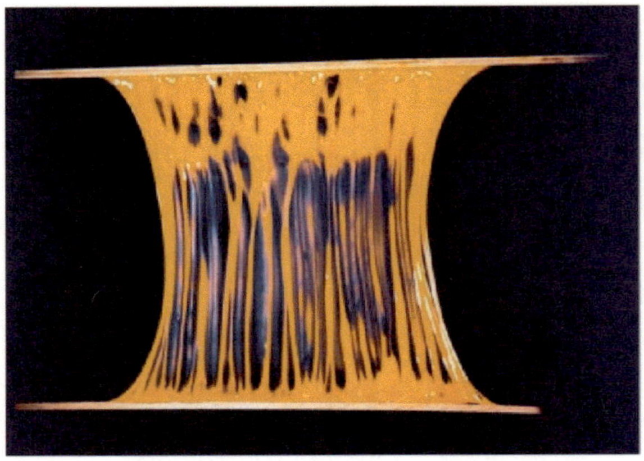

Figure 1.
Viscoelastic nature of pressure-sensitive adhesives.

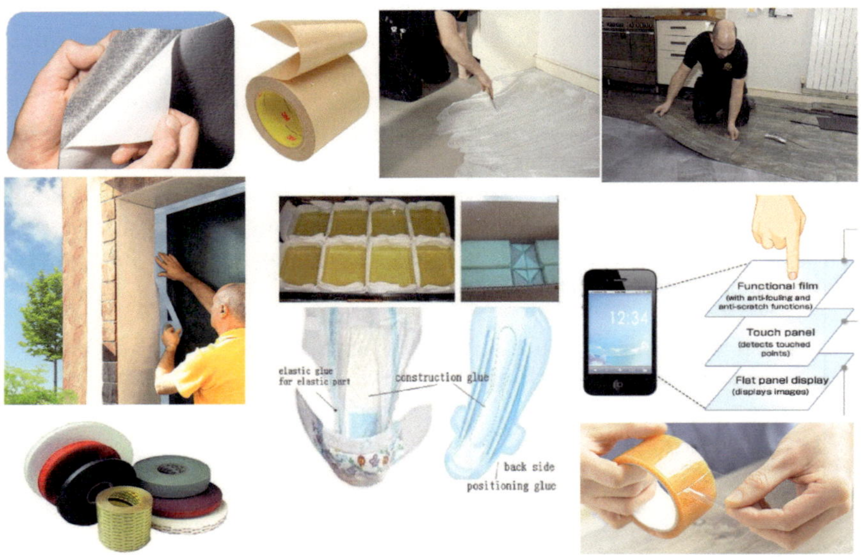

Figure 2.
Various applications of different kinds of pressure-sensitive adhesives.

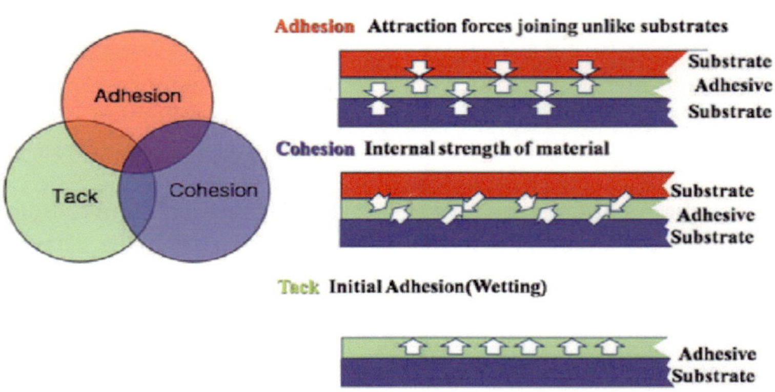

Figure 3.
Major elements of PSA's adhesion properties: tack, adhesion and cohesion.

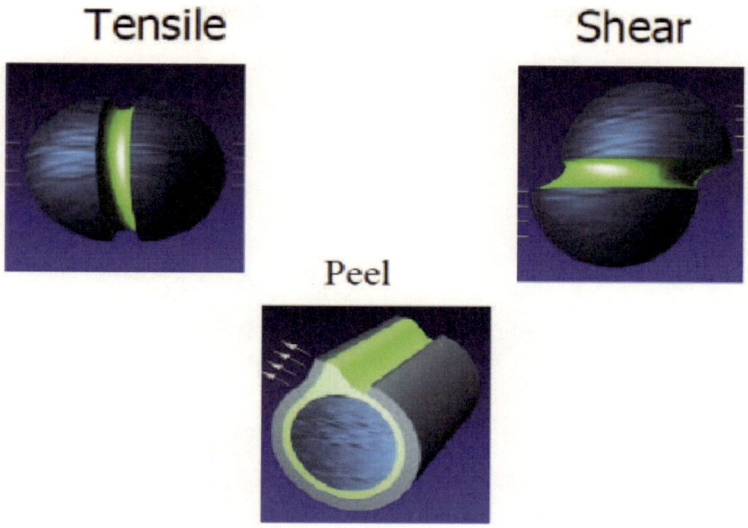

Figure 4.
Schematic of PSA main properties.

When considering pressure-sensitive adhesives and their applications, it's useful to keep in mind main adhesion properties such as shear strength, peel adhesion, tackiness and creep (**Figures 3** and **4**) determine the adhesion of PSA system and subsequently, the application it is used for needs a balance of mentioned properties which can be adjusted via polymerization or formulation process and its components.

In the following chapter, adhesion mechanism of PSAs, types of PSAs, adhesion properties and tests, mechanical and behavior of joints and especially different aspects of PSA applications in the industries will be thoroughly discussed.

2. Adhesion mechanism of PSAs

To understand how PSAs work, it is necessary to understand not only the properties of the individual components of the adhesive formulation and those of the formulation as a whole, but also the processes that take place upon contact of the PSAs with different types of substrates. Good interfacial interaction with a substrate is essential for PSA performance. Such interaction is caused by forces of different natures (van der Waals, H-bonding, acid–base, and donor–acceptor interactions) [2, 4].

At the macro level, this interaction between PSA and substrate results in a wetting– spreading process. Wetting of the substrate by the adhesive is crucial to establish good contact between them. Wetting is also important in the coating procedure (**Figure 5**) [5].

Generally PSAs are used as thin layers; therefore their flow is limited by the physico-mechanical interactions with the solid components of the laminate (liner and face) materials [6, 7]. The term "pressure-sensitive adhesive" itself presumes a definite pressure value necessary to form a strong adhesive joint. Upon converting to dimensionless variables, it would be logical to express pressure as a value relative to the elastic modulus. This is one possible approach to constructing the "dynamic" criterion of pressure sensitivity [4, 8, 9].

The phenomenon of tackiness features a high shear flow of an adhesive material under a compressive force, thus forming the adhesive joint. The flow of the adhesive

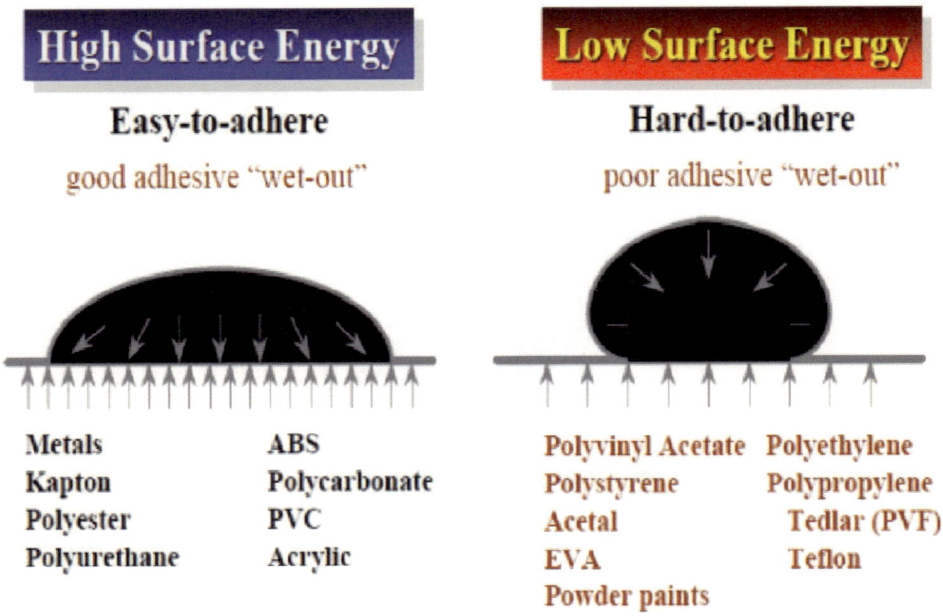

Figure 5.
Wetting of the substrates as the crucial item in adhesion by PSAs.

is needed in the first several seconds to wet the surface of a substrate. Owing to the high viscosity of the PSA, proper wetting cannot be achieved in a reasonable time without compressive force [4].

The final adhesive properties of PSAs require a viscoelastic, non-Newtonian flow behavior which is located at rubbery plateau and related to the supramolecular properties of adhesive. For deeply understanding of viscoelastic behavior, there is the necessity of concluding the most important PSA properties. Totally, adhesives in a linkage have a behavior like a fluid or a solid. PSAs originate from a film-forming, elastomeric material, which combines a high degree of tack with an ability to quickly wet the surface to which it is applied, to provide instant bonding at low-to moderate pressure as a result of its flow characteristics. On the other hand, there is enough cohesion and elasticity in PSAs, in order that in spite of their high tackiness they can be applied with the fingers and easily removed without being left any residue. Furthermore, for achieving bond strength, PSAs must store energy and be elastic. Mainly, there should be a balance between viscous and elastic properties of PSAs. It should also be considered that PSAs have to pass contradictory requirements under different stress rates; at low shear rates, PSAs must flow (bonding) (**Figure 6**) and at high peeling rates there should be an elastic response (debonding) [6, 10–12].

In practical point of view, the relation between adhesion properties and stress rates can be seen by applying peeling test on removable PSAs at different peel rates. The stress rate dependent stiffening is an increase in the elastic contribution to the rheology of the polymer. Major parts of the bond rupture energy are stored when the elastic parts are predominant, causing higher peel and tack properties [13].

Final properties of PSAs are dependent on the nonlinear viscoelastic behavior of the adhesive material being resulted from elastomeric polymer basis of PSAs as viscoelasticity. The same stiffening effect shall be seen by decrease in polymer temperature. This is because the polymer chains are encountered with a restriction in their ability to flow, and the modulus increases. As a result, the adhesive properties of PSAs are also temperature dependent [14].

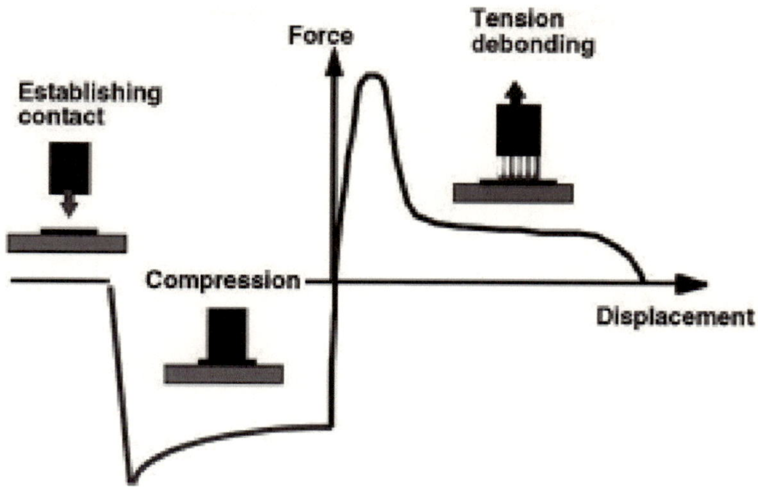

Figure 6.
Schematic of bonding and debonding process.

Pressure-sensitive adhesives must display irreversible work of deformation during bonding and reversible deformation work upon debonding. The ratio of both kinds of deformation work (i.e., of stored and dissipated energy) characterizes the behavior of PSAs.

About the temperature dependence of PSAs' mechanical properties, the storage modulus starts high at low temperatures where all motion within the polymer is frozen and the material behaves like a glass (**Figure 7**).

At higher temperatures it drops off and exhibits a plateau region which represents the elastomeric response generally encountered at normal end-use temperatures; the storage modulus then decreases further when softening begins. The temperature region through which the polymer changes from a glassy (hard) state into a liquid (rubber-like) state, this second order transition point is called the glass transition temperature (Tg), and is a very important parameter in the characterization of PSAs. In temperatures more than Tg, one can apply the time-temperature superposition principle. The side chain size and mobility of the polymer cause different viscoelasticity around Tg [4].

The essential performance characteristics when characterizing the nature of PSAs are tack, peel adhesion, and resistance to shear. The first property represents the adhesive's ability to adhere quickly (initial grab), and latter one evaluates its ability for not being peeled off, and the last one characterizes the resistance against not being flown when encountering with high shear forces. All in all, the first two characteristics are in direct relation with each other, but inversely related to the last one [15].

Tack is measured in two steps according to Zosel: the contact step and the separation one. At first step, in the surface points contact is made, increasing to a larger area while wetting out, viscous flow, and elastic deformation (**Figure 8**) [6].

To be wetted out high fluidity is implied according to the characterization by an appropriate viscosity of the adhesive. Wetting out, being covered the substrate surface by adhesive fluid, results in bonding formation because of the deformability of viscoelastic PSAs. Vice versa, debonding may be assumed the laminate's deformation, the creation and deformation of two new surfaces. Thus, it may be concluded that for bond-forming a high deformation with a medium elasticity is required, while debonding needs an average deformation with a high elasticity [16].

Viscosity and elastic moduli of PSAs as the properties depended on the experimental parameters used, such as the temperature and time, and the strain rate

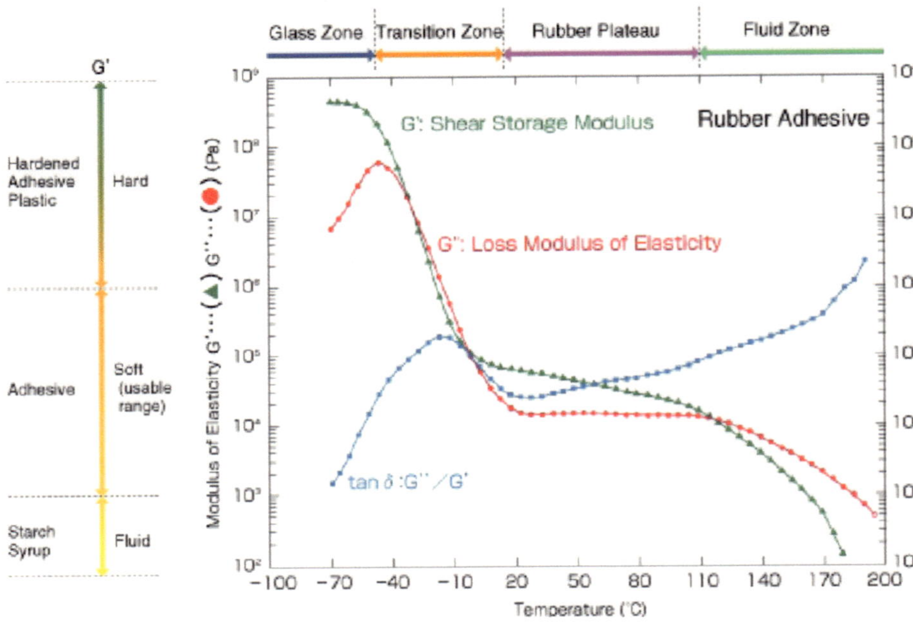

Figure 7.
Temperature dependence of PSAs' mechanical properties.

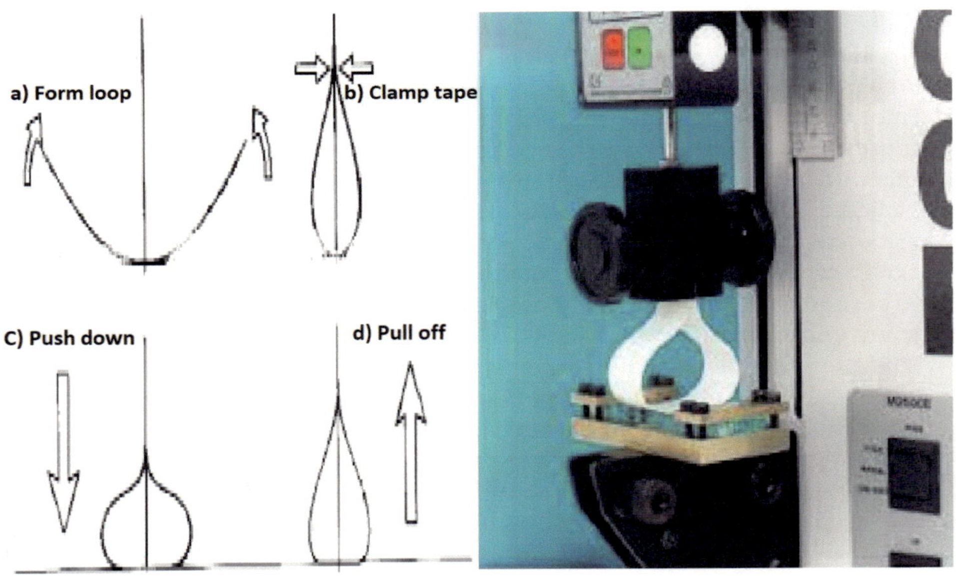

Figure 8.
Contact and separation step schematics in tack test: (a) Form loop (b) Clamp tape (c) Push down (d) Pull off.

cannot be assumed intrinsic material characteristics. So, when speaking of tack, a similar dependence on time, temperature, and the strain rate has to be taken into account which is illustrated by the quite different values of the tack obtained using different experimental techniques namely, quick stick, rolling ball, or loop tack, being evaluated by different time and strain rates, and by the dependence of the adhesive properties to environmental conditions [17].

Peel and peel strength are measured by separating an adhesive applied to a substrate at some angle with respect to the substrate, usually at an angle of 90° or

180°. Similar to tack, the measurement of the peel adhesion involves a bonding step before the debonding or peeling step (**Figure 9**) [7, 15].

Tack as one of the adhesive properties of PSAs is the resistance of the adhesive to being separated after a short time of contact, or by light pressure. Peel is evaluated after an approximately long time of contact after being applied onto the surface of the substrate via a light or medium pressure. The time available for bond forming which means wetting of the surface and then penetration during the first contact step is longer for peeling tests than for tack. After that, flow properties of the adhesive during bonding are less critical for the tack measurement compared to peeling evaluation. Moreover, that how resistant is the PSA in debonding stage is depended on the viscosity/elasticity balance for achieving peelability (removability or repositionability) and on the strain rate influencing the separation resistance substantially than during the measurement of the tack [18–21].

The liquid adhesive must be coated onto a release liner or face material. Good coatability implies adequate machinability or processing properties on the coater (metering roll, drying tunnel) (**Figure 10**) [22].

During production, transport, and coating process, the adhesive fluid is subjected to shear forces resulting in changes of the viscosity. The coated shear-thinned adhesive firstly must wet the web, and the wetting out stage depends on the viscosity of PSA. Except for hot-melt PSAs, the coated fluid adhesive film must go through evaporation of the carrier liquid or solvents (drying) which will result in formation of a solid film. Evaporation of the solvents depends on the fact that how it is diffused into the adhesive layer namely, on the adhesive viscosity. As a conclusion, the viscosity of the adhesive, i.e., the time (shear rate)/temperature dependence of the viscosity, influences the coating process of PSAs [23, 24].

Conventional PSAs can be classified as either nonpermanent (2.7–9.0 N/25 mm for 180_ peel adhesion) or permanent (above 9 N/25 mm for 180_ peel adhesion) (**Figure 11**) [4, 14, 17, 24].

Repositionable PSPs are a special class of removable pressure-sensitive products (labels and tapes) that stick to various surfaces but remove cleanly and can be reapplied. The final adhesion builds up over a few hours. Non-permanent adhesives are

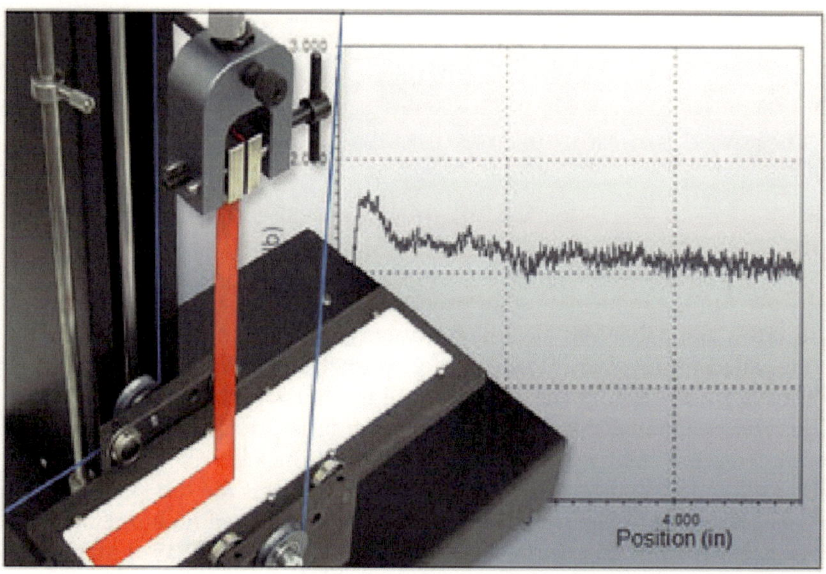

Figure 9.
Peeling test in PSAs.

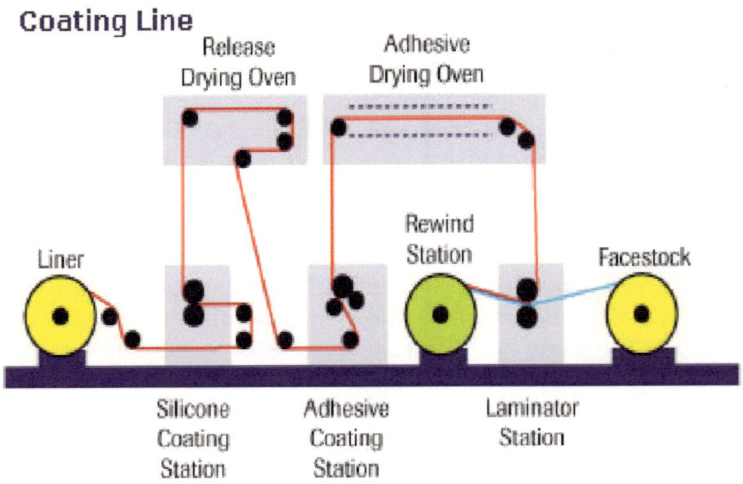

Figure 10.
Coating line of PSAs.

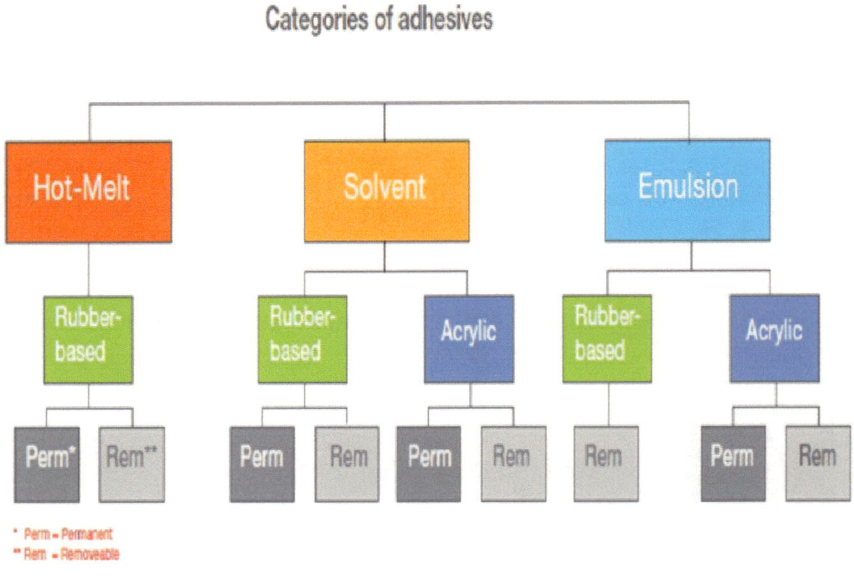

Figure 11.
Categories of adhesives according to the permanency property.

generally used in the manufacturing process of removable tapes and labels, protective laminates, and other less durable products [25].

For so-called removable adhesives (non-permanent adhesives), the flow properties (viscosity), cohesion of the adhesive and the harbor of the adhesive to the face stock are crucial. Ideally, should the bond to the substrate be nonpermanent, a clean separation from the substrate will be expected and the adhesive remains on the face material. One other requirement for good removability is the low peel value with a permanent character. A clean release from the substrate and no build-up of the peel with time are the minimal requirements for removable PSAs. Based on the adhesive properties, it is possible to formulate the rheological properties required for removable and permanent adhesives [1, 6, 14, 18].

3. Types of pressure-sensitive adhesives and applications

Self-adhesive materials are called, in the adhesives trade, .pressure-sensitive adhesives. PSAs are designed to stick on almost any surface by simple contact under light pressure. Among the different classes of adhesives, pressure-sensitive adhesives (PSAs) are perhaps the most common type found in consumer products. Self-adhesive tapes and labels of all kinds are ubiquitous in everyday life. Despite the fact that PSAs are supposed to adhere to substrates, they are different from other adhesives in many ways. First, PSAs are mainly nonstructural adhesives; they cannot be compared with epoxies for structural applications. Moreover, PSAs mainly adhere to a surface upon contact without any chemical reaction [6, 15, 23, 26].

Pressure-sensitive adhesives can be classified by the type of ingredients used to make them (e.g., rubber-based, acrylic and silicone) and by their state (e.g., organic solvent solution, emulsion, solid (hot melt) and aqueous solution) (**Figure 12**).

Emulsions are dispersion of particles containing acrylic polymer in water (latexes). Labels and Packaging Materials (LPM) has manufacturing infrastructures in USA and Europe and wide manufacturing companies to provide latexes around the world. There are also acrylic polymers in petroleum-based solvents—usually of low to moderate solids content. End users prefer the performance of solvent based acrylic polymers although they are more expensive than their water borne counterparts (latexes) [7, 15]. Hot melt PSAs are based on block copolymers. Oils, plasticizers and tackifiers are the necessary components of the formulation needed to obtain appropriate performance. Most hot melts (HM) are dependent on C5/C9 streams from petroleum distilleries. The other PSA acrylic grade UV compositions, which are solventless, produce films with higher thickness, including foam-like adhesives. The two-stage process creating performance advantages in the second stage of curing gives the opportunity of very high molecular weight reactions which can't be cast from solvent. Warm melts (100% solids materials) are either made in solvent and stripped (AC resin) or polymerized to near 100% and coated. They use hot-melt equipment (high coating speeds) and give performance of solvent benefits—high coating speed and solvent-like performance simultaneously.

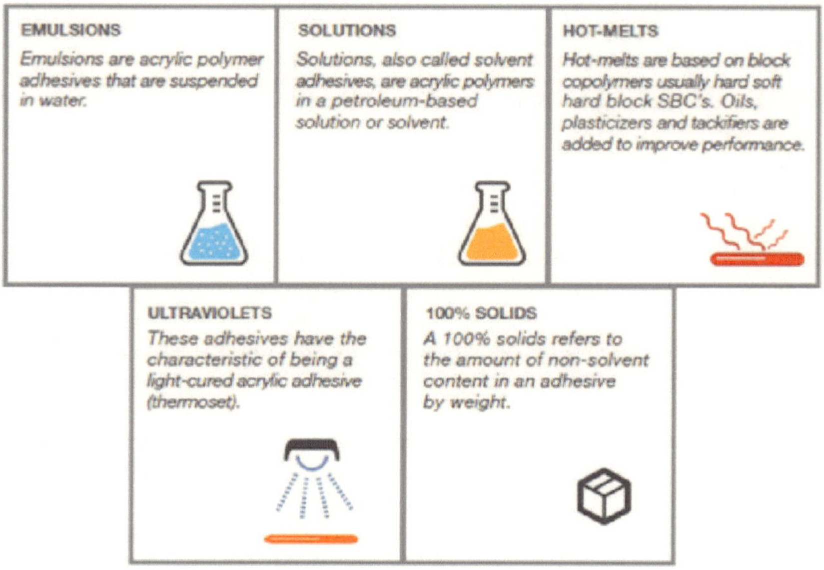

Figure 12.
Different types of PSAs by their state.

PSA are constructed based on these elements as follows:

- **Adhesive:** Typical adhesives are acrylic, natural rubber, or synthetic rubber. Special adhesives with a smaller application base are silicone, butyl, or EVA (ethylene-vinyl-acetate).

- **Backings:** A backing or carrier is a relatively thin, flexible, web based film. Backings can be a paper, film, foil, foam, fabrics, nonwovens, or laminates which are surely chosen based on the application—such as cushioning, dampening, maintaining a barrier, etc.

- **Primer:** A primer is used to increase adhesion of the adhesive to the backing and can be a liquid substance coated onto the backing surface, or a physical treatment, such as corona or flame treatment. This stage is of high importance if the tape needs to be removed cleanly, or if the tape is subjected to stress permanently with a load.

- **Release Coating:** Release coating is a treatment process which should be applied on the backing of a single-sided tape, and the release liner for a double-sided tape. This process ensures stable unwinding of the tape during the application especially for automatic applications in which the tape can be unwound at high speeds. Different release forces can be used for either easy unwinding or harder unwinding, depending on the customer's needs.

Common structure of different tapes and labels are presented in **Figure 13**.

For commercially used PSA systems with excellent performance levels, synthetic polymers based on acrylics, silicones, polyurethanes, and polyesters are preferred (**Figure 14**) [27].

From historical point of view, the first PSAs were rubber based. They are still the cheapest to produce and also the simplest to formulate for they are typically compounds of natural rubber and tackifier which is miscible with the rubber in approximately equal proportions. First versions of rubber based PSAs were not cross-linked, but today a cross-linking step is added to avoid flow [1, 6, 22, 23, 26].

Types of rubber based PSAs are as follows:

- Natural rubber: They must be cross-linked for increasing cohesive strength.

- Reclaimed rubber: This rubber is used to coat the calender surfaces considering its vulcanized properties.

- SBR: This grade must be combined with natural rubber and a small volume of tackifier for enhancement of the adhesion strength.

- Block copolymer: It should be plasticized by heating process via the formation of styrene domains such as SIS and SBS.

- Polyisobutylene: This rubber is a blend of high and low molecular weight resins made by partial crosslinking of butyl rubber and addition of fillers.

Acrylic PSAs are the best choice for optimization and formulation using copolymerization of a long side-chain acrylic such as *n*-butyl acrylate (BA) or 2 ethylhexyl acrylate (2-EHA) with a low glass transition temperature *Tg* (soft monomers), a short side-chain acrylic such as methyl methacrylate (MMA) (hard monomer) to

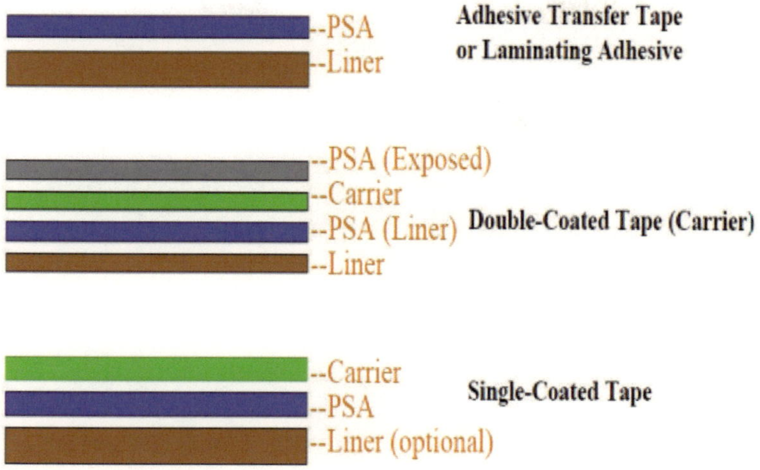

Figure 13.
Common structure of different tapes and labels.

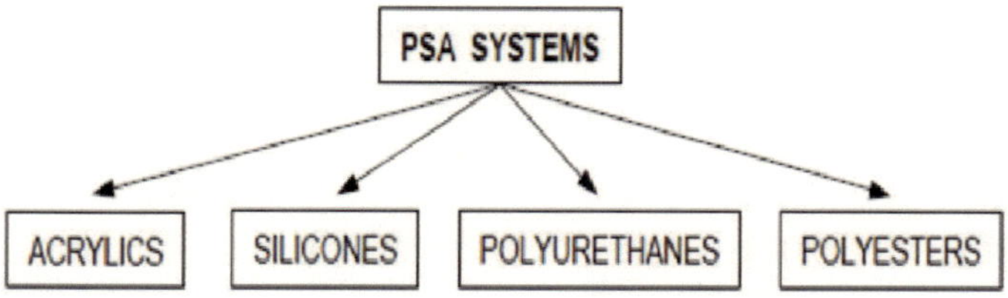

Figure 14.
Commercially used PSA systems.

adjust the *Tg*, and acrylic acid to improve adhesion, optimize elongational properties (their mechanical response to deformation in uniaxial extension) and control colloidal stability [28].

Acrylates and other suitable monomers such as vinyl acetate, acryl amide, and acrylonitrile are copolymerized to synthesize an acrylic copolymer with optimized adhesion properties. Crosslinking agents are also added for cohesion improvement. Acrylics can be synthesized in organic solvents in which no further formulation happens except for the time it is needed to modify their properties. Acrylics can also be synthesized in water (water borne) but here it is needed to add materials called emulsifiers to make the polymer dispersible in water. The third group of acrylics is solvent-free acrylic PSA (**Figure 15**) [7, 15, 27].

Acrylic latex based PSAs are used primarily in the production of different kinds of tapes such as industrial single-sided, double-sided, and transfer adhesive tapes. Another important application of emulsion pressure-sensitive adhesives resides in the manufacture of water-soluble labels. The newest application of them is in the medical products in which neutral electrodes and adhesive tapes are of high importance for securing operating theatre sheets [27, 29–31].

An issue in the field of PSAs is to give them some water solubility depending on the hydrophilic properties of the groups incorporated into the polymer chain. The *Tg* of the water-soluble monomers (**Figure 16**) is the main property for the water solubility and also adhesive properties of the synthesized copolymers [28].

In spite of advantages of water borne acrylic PSAs, the solvent-borne counterparts have been marginally preferred in high-performance applications because of their superior adhesive performance at high temperatures (**Figure 17**). The latter

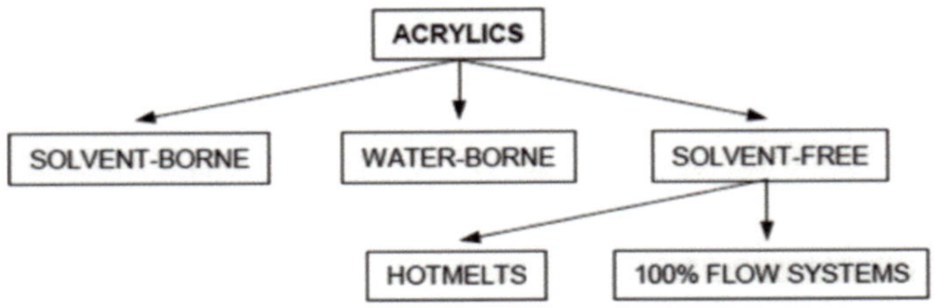

Figure 15.
Acrylic PSAs different types.

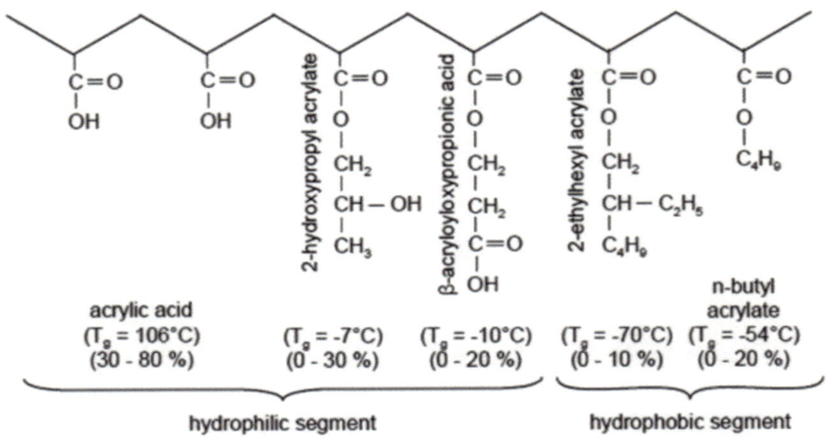

Figure 16.
Structure of emulsion acrylic based PSAs.

one have much higher shear strength, together with higher peel and tack properties, for the aforementioned applications considering the continuous network morphology made by physical entanglements of acrylic chains in the solvent-borne adhesives when the solvent has evaporated [32]. In comparison, the microgels in the latex particles retain their discrete morphology in the adhesive film [33]. This morphology decreases the shear strength of the latex based PSAs because of the poor interconnections between the microgels and the linear polymers in the structure of the film [34, 35].

UV PSAs have the characteristic of being a light-cured acrylic adhesive (thermoset). UV PSAs are nonflammable and solvent free. UV adhesives offer advantages of lower VOCs and better environmental compliance. In the case of saturated acrylic pressure-sensitive adhesives, the photogeneration of initiator radicals by α-cleavage photoinitiators or H-abstraction photoinitiators is followed by a reaction with the acrylic chain that produces a new radical reacting with a neighboring acrylic chain (**Figure 18**) [27].

Totally, acrylic PSA categories are as follows:

Water-soluble

In this grade, monomers having carboxyl groups are copolymerized.

Non–cross-linked

In this one, monomers having low and high Tg are copolymerized to produce a PSA.

Cross-linked

Here, a low Tg monomer is cross-linked by blending with a crosslinking agent.

Figure 17.
Applications of solvent based PSAs such as examples of custom applications for sealing, gasketing, shielding and sound insulation, and foam tapes and adhesives are used in automotive or other industrial applications where strength, heat, or moisture resistance is required (close-up application of heated mirror).

Structural bonding tape (Structural PSAs) is an acrylic pressure-sensitive tape impregnated with an epoxy resin being cured under heat. This adhesive is cured in an oven at about 145°C for 20 min. The composition of structural PSA in which we have inherently pressure-sensitive before the curing process is a combination of soft monomers of acrylate with low *Tg* and hard monomers of epoxy with high *Tg* in the polymer chain (**Figure 19**) [36].

Pressure-sensitive thermosetting adhesive based on acrylic ester/epoxy resin is an initially tacky and conformable thermosetting adhesive produced by blending comprising epoxy resin or a mixture of about 20–60 wt. % epoxy resin and 0, 5–10 wt. % hardener being activated by heating for hardening of the epoxy resins [36].

The other PSA grade which is currently available for industrial application is cross-linkable acrylic hot melt PSAs (HMPSAs) designed to be cross-linked by a UV-lamp or UV-laser light. HMPSAs propose advantages such as relatively low application viscosity at higher temperatures, high processing speeds, aggressive tack, high resistance to plasticizers and solvents, and too high heat resistance after curing process. There are chemically built-in photoreactive functional groups on acrylic chains in UV-acrylic hot melts. UV grades are solvent free and are processed as hot melts. The adhesive film is cross-linked with UV radiation after the acrylic film has been coated with the UV-acrylic hot melt for self-adhesive material production [28, 33].

It is easy to coat heavy depositions of UV grades of acrylic PSA at a good speed since these are 100% solid adhesives. Using photoreactive diluents viscosity can be modified within certain limits and also, the UV dose can be controlled by adjusting the power of UV lamps and the speed at which the substrate is passed under the lamps in the production line.

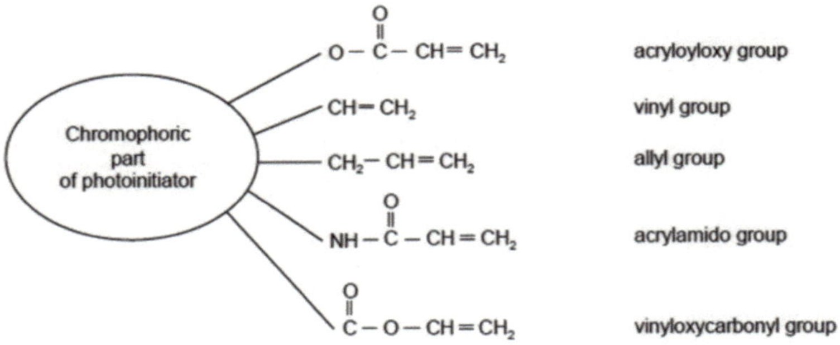

Figure 18.
General examples of the chemical structure of unsaturated photoinitiators [27].

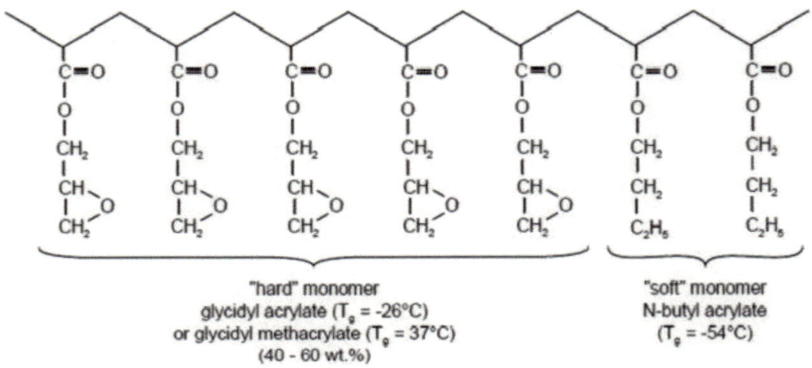

Figure 19.
The polymer morphology of a structural acrylic adhesive.

Pressure-sensitive hot melt adhesives are thermoplastic rubber-based adhesives. They are characterized by the fact that they do not fully set, instead remaining permanently sticky. This allows for an excellent adhesive bond even when the glue is cold. In this case, the required contact pressure is the key to creating a sufficient coat between the contact adhesive and the joined part, which is why this group of adhesives is also referred to as "PSA-pressure-sensitive adhesives" (**Figure 20**).

Styrenic block copolymers based PSAs are the latest type which are introduced into the market. They are generally blends of triblocks of styreneisoprene-styrene (SIS) and styreneisoprene diblocks blended with a low-molecular-weight and high-*Tg* resin based on C5 rings that is miscible with the isoprene phase and not miscible with the styrene phase. It is compulsory to state that a tackifying resin is a necessary component of approximately any class of PSA [26].

Silicone pressure-sensitive adhesives are high-performance adhesives that can be utilized over a wide range of temperatures, from −40 to 300°C (**Figure 21**).

They bond to both low energy and high-energy surfaces. Silicone polymers based on gum contain dimethylsiloxy and diphenylsiloxy groups (**Figure 22**) [37].

The molecular weights of solvent-borne silicones are preferably in the range from 500,000 to 1,500,000 Daltons [38]. The crosslinking of silicone PSAs containing methyl or/and phenyl groups can be achieved thermally between 120 and 150°C using organic peroxide.

With respect to performance, the characteristics of PSA are generally determined by their ingredients. The insert to the right gives a brief summary based on ingredient type [1, 4, 13, 14].

The main applications are as follows:

Temporary bonding (re-bonding not required; easy removal preferred).

Bonding “hard to bond adhered (e.g., PP, PE, PTFE, and EPDM).

Temporary bonds (e.g., adhesive processing of nameplates, adhesive coatings on labels, films).

Primary processing of adhesive components (e.g., double-sided adhesive tape).

Repeated peeling and adhesion (requiring enhanced re-adhesion and peeling properties).

It’s important to select the appropriate type of PSA for your needs.

Figure 20.
PSA hot melt.

Figure 21.
Silicone PSA application in electronic circuits.

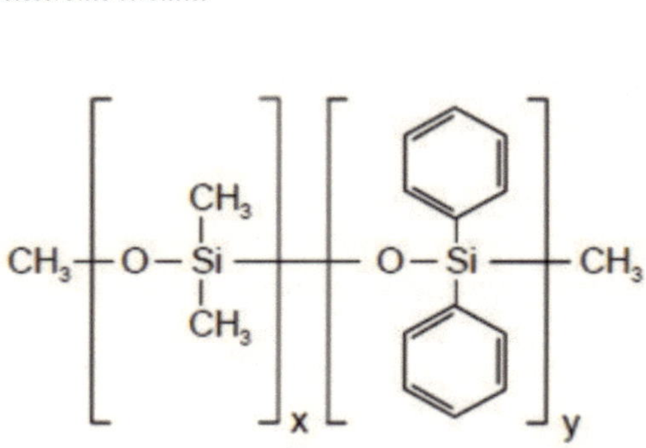

Figure 22.
Silicone PSA structure.

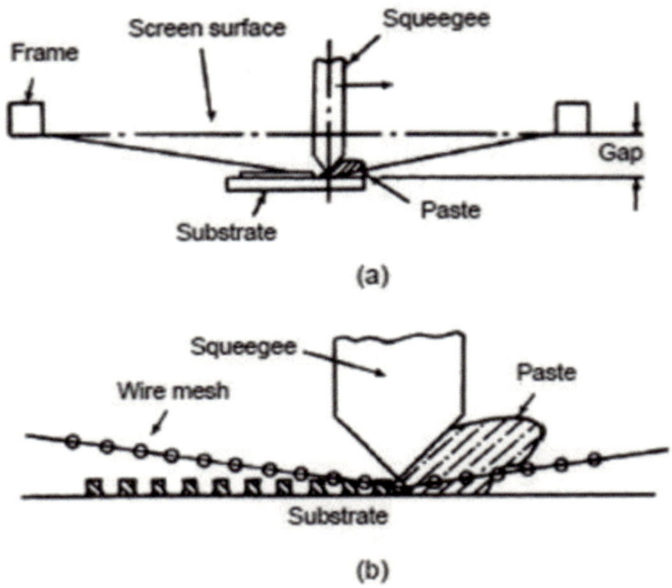

Figure 23.
Screen printing: (a) Pattern making (b) Printing.

Screen printing which is one application of PSAs initiates with stretching a piece of screen (mesh) over a frame and securing it into place. After that, a photoreactive emulsion that cures when light is irradiated onto it is applied to the mesh. Then, the desired pattern is placed onto the emulsion-coated mesh (**Figure 23**). In the next stage, the sealant emulsion was used to form a resist when being exposed to light. To remove the unhardened emulsion, the screen is washed, and the pattern emerges as an opening not covered by the resin which can be used as the plate for screen printing. For printing, ink is then placed inside the frame, and the plate is placed over the substrate to be printed. The pattern is transcribed onto the substrate when the ink is pushed out of the opening with the linear pressure from the squeegee. This process can be repeated many times to mass produce substrates with the same pattern [39].

Before the film assembly on LCD, other processes including punching, transporting and inspecting are necessary. Thus, the application of a protective film is compulsory to protect the surface of the optical films from contaminants or scratching. A sketch of a polarizing plate with a protective film is demonstrated in **Figure 24**. The substrates of the protective film are typically fabricated using polyethylene (PE) or polyethylene terephthalate (PET) with pressure-sensitive adhesives (PSAs) coated on one side. PSAs are viscoelastic materials that can adhere to various substrates upon application of light contact pressure in short contact time [40].

There are two different major classes of medical adhesives available [41]:

- Permanent or tissue adhesives (glue).
- Removable adhesives (pressure-sensitive adhesives).
- Other than acrylics, there are other types of PSAs for medical applications:
- Silicone-based adhesives, which are used in transdermal drug delivery systems (drug delivery). These devices are in need of an inert adhesive and biocompatible system.

- Polyvinyl ether-based adhesive, which are used in permeable skin patches.
- Polyvinyl pyrrolidone-based adhesives, which are used for high moisture absorption applications, such as ostomy bags and mounts, without the loss of adhesion.
- Urethane-based adhesives, which are used for sustained released wound dressings and urethane implants.

The biggest amount of the pressure-sensitive adhesives on the market is for wound protection. The applications for PSAs are:

- Transdermal drug delivery.
- Adhesive bandage or plasters.
- Adhesive drapes.
- Ostomy bag mounts: construction of an opening in the human skin. For example, a tracheostomy.
- Intravenous needle attachments, like morphine perfusion.
- Electro medical procedures.
- Microporous tapes.

Drug-loaded adhesive patches delivering the therapeutic agent, when applied to the skin, at a controlled rate, through the skin to the systemic circulation and to the target organs are transdermal drug delivery (TDD) systems. For maintaining intimate contact between the patch and the skin surface, adhesives are used in TDD applications. Many types of adhesives might be considered to be used for application of TDD patches although PSAs are preferred. There are other factors to be considered in the selection and use of PSAs for TDD systems in addition to the usual properties of PSAs such as tack, adhesive and cohesive strength including compatibility, biocompatibility, delivery system compatibility, formulation, and acceptable regulatory status being equivalent supporting documentation or the availability of drug master files [42].

The adhesive matrix or drug-in-adhesive (DIA) design in its simplest form is directly applied into the PSA matrix. DIA's design configuration is shown in **Figure 25a**.

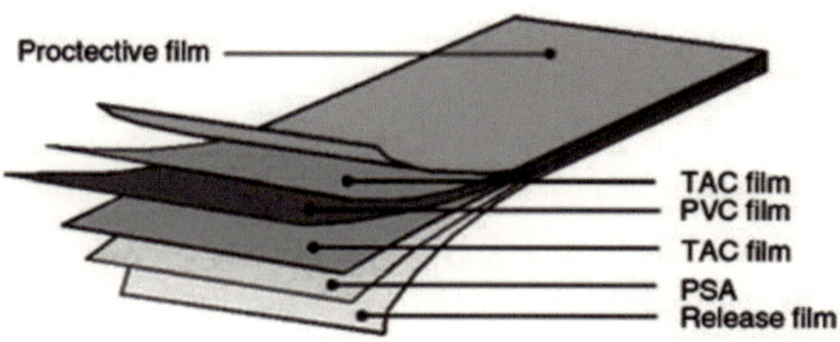

Figure 24.
A polarizing plate with a protective film.

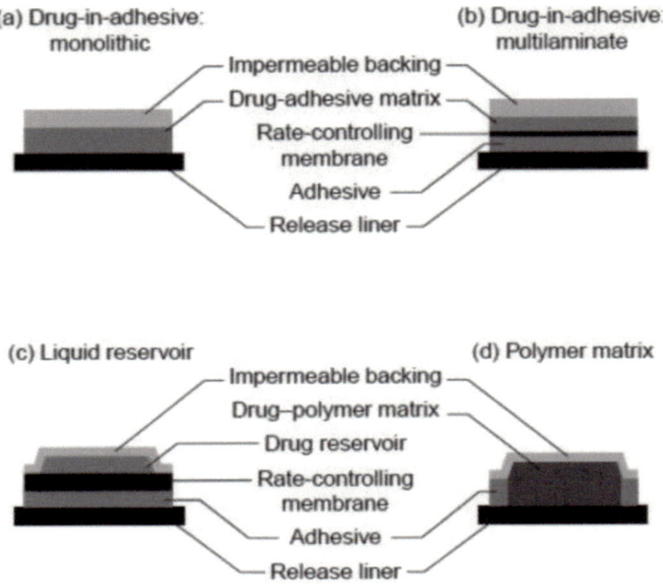

Figure 25.
Typical designs of TDD systems.

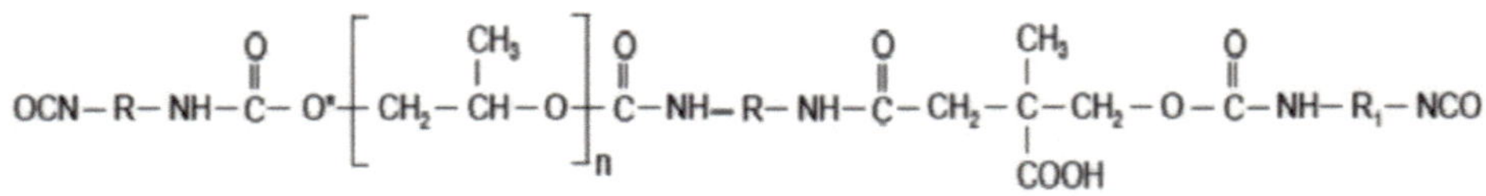

Figure 26.
Structure of polyurethane based PSA.

The adhesive matrix transfers several properties, including peel adhesion, control over drug delivery rate and storage of the drug, and it also governs their partitioning into the stratum corneum. An impermeable backing film supports DIA matrix on the topside and on the side facing the skin there is a lamination with a removable release liner [1, 42].

There is also a contribution of the extraordinarily diverse chemistry of polyurethanes to the development of efficient adhesives applying only to a limited degree to PSAs for which the number of patent publications is surprisingly small. Solvent free 2-K PSAs on the basis of polyether polyols and isocyanates are available on the market (**Figure 26**) [27].

Previous studies have shown [43] that aliphatic polyesters with molecular weights ranging from 10,000 to 50,000 Dalton and *Tg* between –25 and – 47°C can give us excellent adhesion. An example of this method is the acrylation of polyester PSAs containing end groups of hydroxyl. Introduction of an unsaturated architecture can also be practically achieved via the reaction of the OH-groups of a polyester chain with the corresponding acrylic or unsaturated isocyanate compounds [44].

4. Manufacturing of PSAs

The manufacture of pressure-sensitive raw materials and adhesives can be carried out off -line or in-line. As illustrated in **Figure 27**, both technological modalities include various chemical technologies. Generally, off -line synthesis yields pressure-sensitive raw materials or PSAs; in-line technology leads to simultaneous PSA and PSP (pressure-sensitive products) manufacture [24].

As illustrated in **Figure 28**, off-line synthesis (1) allows polymer-based PSA formulation to be coated (2) and (if necessary) transforms (mainly by cross-linking) in PSP.

Pressure-sensitive products (PSPs) are multilayer constructions based on solid state materials and coated component. Therefore, the production of the PSPs includes the manufacture and assembly of the individual components. The manufacture of the components includes the manufacture of both the solid-state carrier materials (face stock and release liner) and the coating components (adhesive, adhesive, primer, etc.) Generally, the buildup of the PSP from its components is carried out by coating and laminating and leads to a web-like product that must be confectioned. **Table 1** presents the manufacturing technology of main PSPs and demonstrates that the manufacture of PSPs by coating of a solid-state carrier with a liquid pressure-sensitive adhesive (PSA) is the main procedure to produce various

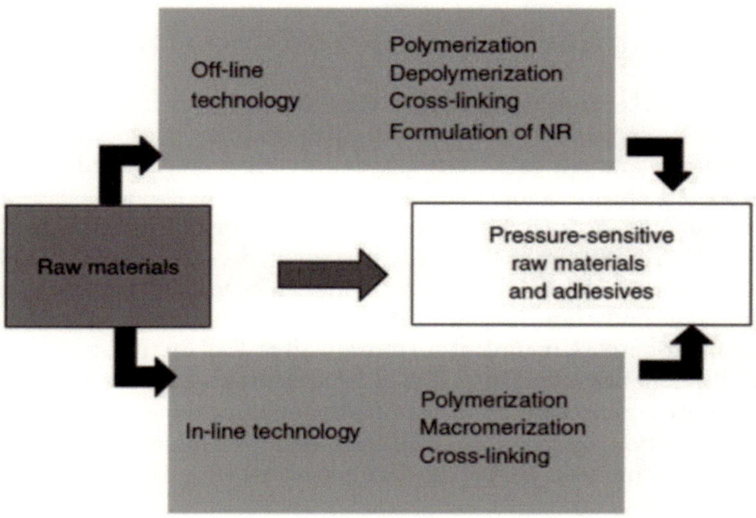

Figure 27.
Principles of manufacture technology for pressure-sensitive raw materials and adhesives [24].

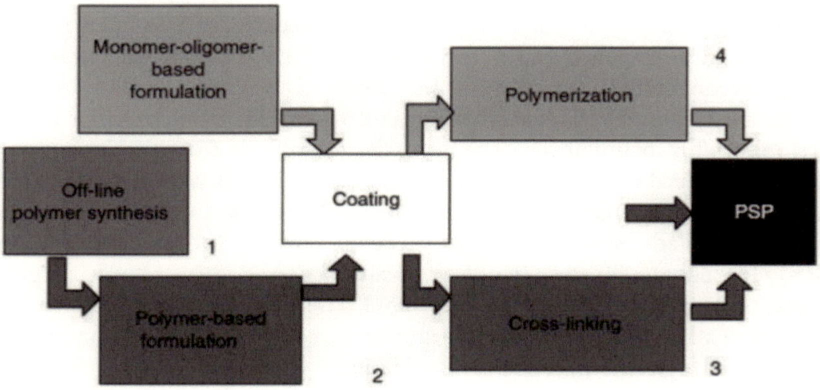

Figure 28.
Manufacturing steps for PSPs [24].

Manufacturing procedure	**Parameter**						
	Free choice of product components		Running speed			On-line manufacture	PSP
	Carrier	**Adhesive**	**Very good**	**Good**	**Fair**		
Coating	●	●	●			●○	Label, tape, protection film
Extrusion coating	--	--		●		●○	Tape, protection film
Coextrusion	--	--			●	●○	Tape, protection film

Note: ●, preferred; ●○, possible.

Table 1.
Manufacturing technology of main PSPs [24].

PSPs. The principal coating component is the PSA. Other coating components are used to allow the anchorage of the adhesive on the carrier (e.g., primer) and its debonding from the laminate (e.g., release agent) and to protect or finish the adhesive or the carrier (e.g., printing inks, lacquers) [1, 6, 17, 24].

PSAs manufacturing includes the manufacture of the raw materials and also their formulation. PSAs can be found in different forms such as solutions, dispersions, or hot-melt adhesives. Water-borne PSAs can be used in certain cases without any additional formulation. Solvent-based acrylic PSAs may also be used without further formulation. PSA manufacture is originally based on the blending and dissolving of different raw materials, none of which exhibits a PSA character lonely.

For preparation recipes for PSAs, formulating must be done via changing the adhesive composition, by addition of different macro- or micromolecular compounds with elastomeric or viscous, surface active or chemically stabilizing properties; antimicrobial agents must also be added to the adhesive. PSAs should have a Tg of about _15° to _5°C. Latexes' sensitivity to bacteria resides in the moderate pH and presence of polyvinyl alcohol. Blending a high Tg, hard resin with a lower Tg elastomer provides a desirable formulation. Considering labels, PSA coating and converting, and end-use require other special performance requirements which can be obtained only through formulating. For the formulation of PSAs, it is needed to produce tailored adhesives for each one of the intended applications. Parameters including the converting technology of the adhesive, the coating technology, the end-use properties, and economical criteria influence the formulation of PSAs. The first reason for formulating is the necessity to improve the adhesive properties. Rubber based PSAs do not possess appropriate intrinsic tack and peel adhesion. Most viscoelastic materials with intrinsic tack being obtained via the synthesis do not show enough peel adhesion to nonpolar surfaces, and soft, tacky PSA raw materials exhibit too low shear adhesion. In cases of water borne or solvent borne acrylic PSAs, the removability of the PSA must be improved. Therefore, a proper balance between high tack, peel adhesion, and high cohesion has to be achieved using formulation. Formulating for high tack and peel (tackification), using viscous formulating agents such as tackifiers and plasticizers decreases the shear adhesion. A high cohesion level may be achieved by the rubber-like component of the adhesive or by partial crosslinking [6, 15, 22, 24].

There are rubber-like (elastic) and liquid like (viscous) PSA components.

Natural raw materials with rubber-like properties and natural resins or plasticizers may be used in order to formulate an adequate recipe. Physical modification allows the use of these materials in a suitable state [22, 27].

Natural rubber is used as a solution, and manufacture of the PSAs implies dissolution of the rubber. In fact a slight mechanochemical destruction of the rubber occurs in parallel with the dissolution process. At least theoretically a precalendering (i.e., a controlled depolymerization of the natural rubber) is possible, but not economically viable. There are only a few special uses (like protective films) where mastication of the rubber must be carried out. Masticated rubber was used for tapes. Pressure-sensitive adhesives exhibiting low tack, for bonding onto themselves, also are formulated on the basis of masticated natural rubber [4–4]. The high molecular weight portion of natural rubber is insoluble in solvents and therefore natural rubbers must be milled to a Mooney viscosity of 53–75 or below, to obtain solubility at 10–30% solids in solvent [6].

Solvent-based PSAs are tackified elastomers, tackified or untackified viscoelastomers, or tackified plastomers. Both (the tackified and the untackified recipe) need adhesive solutions as a blending medium; that is, mixing of the components occurs in solvents. Classic, solvent-based adhesives were formulated with natural rubber (NR) and tackifier resins. Later, synthetic rubber-like products were introduced. Acrylics, vinyl acetate (VAc) copolymers, silicones, polyurethanes (PURs), and other viscoelastomers enlarged the choice of raw materials for solvent-based PSAs and the diversity of the required equipment. Solvent-based acrylics need a special design, but less formulation. It is well known that various materials can be used as adhesives [45–47], among which acrylic polymers are the most popular ones for PSA applications because of the inherent unique properties. Acrylic PSAs, generally borne in an organic solvent, must possess correspondingly suitable 180° peel strength, high shear strength and appropriate tack for optical film applications. The adhesive properties of acrylic PSAs are critically determined by the monomer compositions, the initiators, the crosslinking agents and the tackifiers of acrylic PSAs [48–50].

The main differences between solvent-based and water-based systems are that solvent based systems are homogeneous and water-based ones are heterogeneous; solvent-based systems have low molecular weight (limited by the solubility/viscosity of macromolecular compounds), whereas water-based ones possess higher molecular weight; solvent based systems may be formulated without non adhesive components, but water-based systems must include technological additive. Started with rubber–resin PSA solutions, label coating now uses mostly acrylic, water-based PSAs. Therefore, the formulation of PSAs for labels is mainly a formulation of water-based compounds. In practice, it consists of the blending of liquid-state components, that is, the viscoelastomer dispersion and the tackifier dispersion. It should be taken into account that in a different manner from true solutions, water-based dispersions have a limited (4-year) shelf life, narrow storage-temperature intervallum (−30 to +70°C), and low shear stability. First step of the manufacturing process is the emulsion polymerization process of acrylic monomers in a polymerization reactor plant. After that, this water borne acrylic polymer solution called latex has to be mixed with other additives to be formulated as a PSA applicable for PSP applications. Blending acrylic dispersions requires a test of compatibility, adjustment of pH, and, after the addition of solvent, plasticizers, or fillers, storage time. Peel improvement after longer storage times for a tackified acrylic dispersion is also necessary [1, 6, 7, 15, 51, 52].

Hot melts are based on special designed polymers that need complex formulation. Generally, hot-melt PSAs (HMPSAs) are formulated from thermoplastic elastomers (TPEs) and viscous (liquid) components. The manufacture of HMPSAs requires mechanical size reduction, micronization of certain elastomers (TPEs are generally pelletized), and molten state mixing of them with other components of the recipe (tackifiers, oils, antioxidants, etc.). Such mixing is carried out discontinuously or continuously [6] with a special apparatus that produces a heat history of the material. For continuous mixers the heat history is very low. Principally, melt mixing can be carried out using a variety of heated mixing equipment. In order of increasing effectiveness, these include vessels stirred with propellers, vessels with high shear (Cowles type) dispersers, planetary mixers, internal mixers (Sigma blade or Banbury type), and modifications of these. In low shear mixers, resins, plasticizers, and stabilizers should be melted first and thermoplastic rubber crumb added incrementally. Where production rates are high enough, continuous mixers are desirable because they reduce degradation of the rubbery mid-block. For hot-melt manufacture, compounding is mixing of the adhesive components at 120 and 200°C, where the shear forces cause a drop in the viscosity. The construction and hydrodynamic characteristics of the mixing equipment may influence the processing performances of HMPSAs; the raw materials used affect it also. For instance, when polyacrylate rubber-based HMPSAs are prepared, high shear mixing is required. As a rule of thumb, increasing mixer shear and temperature lead to better mixing of the adhesive components, a breakdown in molecular weight, and a subsequent increase in tack and reduction in shear performance. In some cases, these effects may be signify cant enough to overwhelm any effects caused by changing the adhesive formulation; for that reason, some researchers choose to solvent blend and solvent coat samples rather than use a mechanical mixer or extruder. Furthermore, adhesive melt viscosity and shear strength are extremely sensitive to melt processing and subsequent breakdown in molecular weight, whereas tack and peel resistance are relatively robust and will remain relatively unchanged [22, 24].

Generally, the manufacture of PSPs includes the manufacture of solid-state components (e.g., carrier materials for face stock and for release liner), the manufacture of liquid adhesive and nonadhesive components (e.g., PSA, release and primer), and their assembly in a continuous web-like material. The PSA can be coated on or built into the bulk carrier material. The classic way to manufacture a PSP is by coating a carrier material with a PSA. Manufacture of the finished product by carrier coating employs a coating technology completed by other converting steps that are common for web-like products. Web finishing is the coating of liquid product components on solid-state web components or their build into solid-state web components. From the technological point of view, PSPs can be manufactured using coating, extrusion coating, or coextrusion. Generally, coated adhesives are components that are deposited on the surface of a carrier material. The manufacture of PSPs by coating is a common procedure for classic products. Coating technology uses various coating methods; their choice depends on the physical state of the adhesive (e.g., solvent-based, water-based, or HMPSA) and PSP grade. For labels, roll-coating and screen printing offer the best results, but blade, cast, and slot-die coating are used also. For tapes, roll-coating, slot-die coating, extrusion, and calendering are mostly used, but blade, cast, and spray coating were suggested as well. Protection films are manufactured by roll and blade coating. Different protective are illustrated in **Figure 29** [22, 24, 28].

Coating methods can be classified as impact (i.e., contact) coating (e.g., roll-coating, screen printing) and nonimpact coating (e.g., spray coating, curtain coating) and according to the dosage of the coated material, that is, whether it is in excess or metered (**Table 2**).

There is also manufacturing of PSPs via extrusion process. Actually, pressure-sensitivity of PSPs resides in their coated or built-in pressure-sensitive layer. Buildup

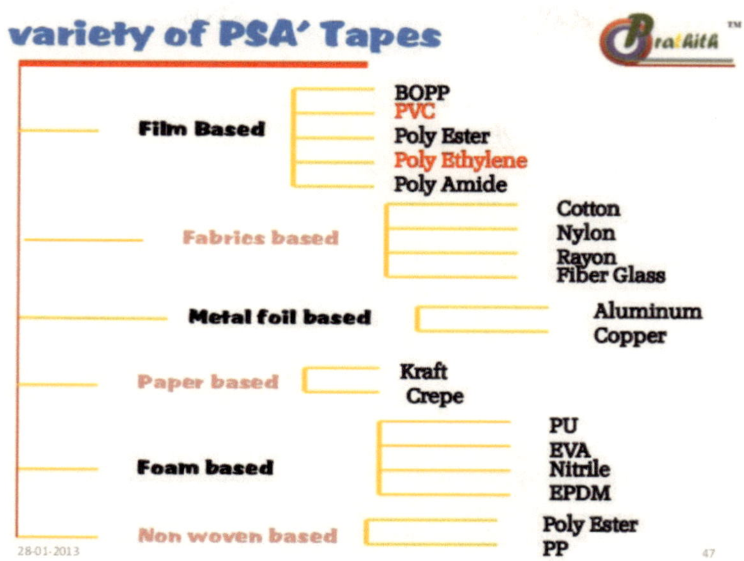

Figure 29.
Different PSA tapes based on protective films.

Impact coating		Nonimpact coating	
With adhesive excess	**With adhesive dosage**	**With adhesive excess**	**With adhesive dosage**
Knife coating	Slot-die[c]	Air brush[f]	Spray coating[d,f]
Roll coating[a]	–	–	Curtain coating[e]
Screen coating[a]	–	–	
Calendaring[b,c]	–	–	–
Main use in other technologies[c]			

[a]*Printing (gravure, flexo).*
[b]*Rubber processing.*
[c]*Plastic film manufacture (cast, chill-roll film).*
[d]*Printing (ink jet).*
[e]*Plastic film manufacture (plasticized cast film).*
[f]*Lacquering.*

Table 2.
Working principle of different coating devices [24].

of a PSP from its components means the classic manufacture of the finished product which are the carrier material and the PSA, respectively. Practically, other product components such as the release liner and cover film can be built in through one or more layers resulting in a laminate with a complex structure. The PSP can also be manufactured in one step, offering primarily economic advantages if a carrier material with built-in pressure sensitivity is used. As a result, economic considerations forced the development of extrusion-made products. **Table 3** concludes the main properties of the manufacture of adhesiveless and adhesive coated PSPs [22].

When manufacturing PSPs, the final properties of the product and trends in production technology must also be considered. The selection of the PSP components in the manufacturing process of it depends on the product type. There are some limitation in each product class regarding the choice of manufacture equipment and

Raw materials	Manufacturing procedure	PSP
Polyethylene	Extrusion	Protective film
Ethylene-vinyl acetate copolymers	Extrusion, coextrusion	Protective film, business forms
Ethylene-olefin copolymers	Extrusion	Protective film, decoration film
Ethylene-acrylate copolymers	Coextrusion	Protective film, tape
Butene-isobutene copolymers	Coextrusion	Protective film, tape
Vinyl acetate copolymers	Extrusion	Business forms
Vinyl acetate copolymers	Extrusion, casting	Tape

Table 3.
Major raw materials and manufacturing processes for PSPs [22].

laminate components namely printing technology consideration in label manufacture which determines the choice of the label components and manufacturing method. The manufacture of labels should begin with the design process and also include digital artwork and reproduction, origination and pre-press electronic design and artwork plate types, and plate making,. Changes in technology go back to the material components or new technologies used in production. The printing industry has undergone changes to meet these standards. Water-based, solvent-free, and 100% solids adhesives are being used more and more to meet the law of the toxic products emission standards. Food laws get tougher in controlling the effect of migration of various components of the package and also their interaction with the packed products. These changes motivate the use of solvent-free and adhesive-free products.

5. Conclusions

As wholly discussed in this chapter, PSAs as viscoelastic, non-Newtonian flow behavior which is based on the macromolecular nature of the adhesive are increasingly replacing more traditional adhesives because of their ease of use and safe manipulation. Environmental limiting laws drive research toward other production methods to coating from solution in spite of the developments of energy-efficient solvent-recovery methods. Acrylic latexes are so popular for general-use PSAs. However, there is a limitation for their access to the most demanding applications due to the presence of surfactants in their microstructure, which are not easy to eliminate and the lesser control over the molecular structure. The manufacture of pressure-sensitive raw materials and adhesives can be carried out off -line or in-line. An increasingly popular production method is hot-melt, or solventless, technologies. In this process, the viscosity should be lowered enough for an easy coating without the use of any volatile compound. Regarding end-use properties, the future trend for PSAs will be in the direction of heterogeneous polymer structures and also on the incorporation of additional functionalities such as thermal or electrical conductivity or controlled drug release.

Author details

Jaber Khanjani
Department of Chemical, Materials and Polymer Engineering,
Buein Zahra Technical University, Buein Zahra, Qazvin, Iran

*Address all correspondence to: jaber.khanjani@gmail.com

References

[1] Eastman EF, Fullhart L. Handbook of Adhesion. 1998. Chapter 23

[2] Satas D. Handbook of Pressure Sensitive Adhesive Technology. 2nd ed. New York: Van Nostrand Reinhold; 1986

[3] Dillard DA, Pocius AV, Chaudhury M. The Mechanism of Adhesion. 1st ed. Elsevier B.V.; 2000. p. 20

[4] Benedek I, Feldstein MM. Handbook of Pressure Sensitive Adhesives and Products. USA: Taylor & Francis Group, LLC; 2009

[5] Anderson C. Pressure Sensitive Adhesive Articles and Method of Making Same. US Patent 3,769,254. 1973

[6] Benedek I. Pressure Sensitive Adhesives and Applications. New York, NY: Marcel Dekker Inc; 2004

[7] Khanjani J, Zohuri GH, Gholami M. 11th International Seminar on Polymer Science and Technology. Tehran, Iran: Iran Polymer and Petrochemical Institute; 2012

[8] Brooks TW, Kell RM, Boss LG, Nordhaus DE. In: TAPPI Proceedings. TAPPI Polymers, Laminations and Coatings Conference. Boston. 1984. p. 469

[9] Mablitt DF. Engineer's Handbook of Adhesives. 1st ed. Great Britain: Whitefriars press. Ltd.; 1972

[10] Moghbeli MR, Zamir SM, Molaee B. Resultant synergism in the shear resistance of acrylic pressure-sensitive adhesives prepared by emulsion polymerization of n-butyl acrylate/2-ethyl hexyl acrylate/acrylic acid. Wiley Inter Science. 2007;**57**:1226-1236

[11] Shen H, Zhang J, Gin S, Gudongo L, Zhang L, Qu X. Usability testing of water-soluble pressure-sensitive adhesives to labels. Prosimy cytować jako: Inżynieria i Aparatura Chemiczna. 2007;**68**:1354-1365

[12] Antosik AK. Czech Z. Pressure-Sensitive Adhesives (PSA) Based on Silicone. book-chapter. Wiley. http://dx.doi.org/10.1002/9781119242604.ch7

[13] Czeck Z, Kowalczyk A, Swiderska J. Pressure-Sensitive Adhesives for Medical Applications. Szczecin, Poland: West Pomeranian University of Technology; 2011. Available from: www.intechopen.com

[14] Severtson S, Guo J, Xu H. U.S. Department of Energy Project Number DE-FC36-04GO14309

[15] Khanjani J, Zohuri GH, Gholami M, Shojaei B, Dalir R. Emulsion semi-batch terpolymerization process using hybrid emulsifiers for synthesizing new emulsion pressure sensitive adhesives (EPSAs). The Journal of Adhesion. 2014;**90**:174-194

[16] Zosel A. Adhesive failure and deformation behaviour of polymers. Journal of Adhesion. 1989;**30**:135

[17] Chaudhury M, Pocius AV. Adhesion Science and Engnieering. Vol. 2. Amsterdam, The Netherlands: Elsevier; 2002

[18] Chum CM, Ling MC, Vargas RR. High Performance Adhesives from Ethylene Copolymers via Grafting. Avery Int. Co.: USA; 1987. EP 12257927/18.08.87

[19] Rana PK, Sahoo PK. Synthesis and pressure sensitive adhesive performance of poly (EHA-co-AA)/silicate nanocomposite used in transdermal drug delivery. Journal of Applied Polymer Science. 2007;**6**(106):3915-3921

[20] Aymonier A, Leclercq D, Tordjeman P, Papon E, Villenave JJ. Control

of structure and tack properties of acrylic pressure-sensitive adhesives designed by a polymerization process. Journal of Applied Polymer Sience. 2003;**89**:2749-2759

[21] Tobing S, klein A, Sperling LH, Petraske B. Molecular parameters and their relation to the adhesive performance of acrylic pressure-sensitive adhesives. Journal of Applied Polymer Science. 2001;**81**:2109-2117

[22] Benedek I. Developments in Pressure Sensitive Products. USA: CRC Press, Taylor & Francis Group, LLC; 2006

[23] Petrie EM. Handbook of Adhesives and Sealants. USA: McGraw-Hill Companies, Inc.; 2000

[24] Benedek I, Feldstein MM. Technology of Pressure Sensitive Adhesives and Products. USA: Taylor & Francis Group, LLC; 2009

[25] Johnson M, Hattam P, Purgett MD, Everaerts AI, Kczorek K. US Patent. 1998

[26] Creton C. Pressure sensitive adhesives: An introductory course. MRS Bulletin. 2003 Available from: www.mrs.org/publications/bulletin

[27] Czech Z, Milker R. Development trends in pressure-sensitive adhesive systems. Materials Science-Poland. 2005;**23**(4):1015-1022

[28] Lakrout H, Creton C, Ahn D, Shull KR. Influence of macromolecular feartures on the tackiness of acrylic polymer melts. Macromolecules. 2001;**34**:7448

[29] Bernard MM. US Patent 5,623,011. 1997

[30] Kazuhiro, M. US Patent 4,826,907. 1989

[31] Ishiava K. US Patent 6,190,767. 2001

[32] Tobing SD, Kelin A. Molecular parameters nd their relation to the adhesive performance of acrylic pressure-sensitive adhesives. Journal of Applied Polymer Science. 2001;**79**:2230

[33] Jovanovic R, Dube MA. Screening experiments for butyl acrylate/ vinyl acetate pressure sensitive adhesives. Industrial and Engineering Chemistry Research. 2005;**44**:6668-6675

[34] Aubery DW, Ginosatis S. Peel adhesion behavior of carboxylic elastomers. Journal of Adhesion. 1981;**12**:189

[35] Mao TJ, Reegen SL. In: Weiss P, editor. Proceedings of the Symposium on Adhesion and Cohesion. Amsterdam: Elsevier; 1962. p. 209

[36] Czech Z, Urbala M, Martysz D. Investigation of the tensile properties of “epoxy-resin-monomer” systems using cationic UV-crosslinkable blends. Polymers for Advanced Technologies. 2004;**15**:387

[37] Czech Z, Milker R. Developments trends in pressure sensitive systems. Materials Science. 2005;**23**(4):1015-1022

[38] Mazurek M. US Patent 5,314,748. 1993

[39] Three Bond Technical News. Japan. 23 June 1992

[40] Jin X, Bai YP, Shao L, Yang BH, Tang YP. Properties of solvent-borne acrylic pressure-sensitive adhesives synthesized by a simple approach. Express Polymer Letters. 2009;**3**(12):814-820

[41] Piau JM, Verdier C, Benyahia L. Influence of rheology and surface properties in the adhesion of uncross-linked pressure sensitive adhesives. Rheologica Acta. 1997;**36**(4):449-461

[42] Tan HS, Pfister WR. Pressure-sensitive adhesives for transdermal

drug delivery systems. Pharmaceutical Science & Technology Today. 1999;**2**(2):60-69

[43] Acevado M, Huber H. Proceedings RadTech Europe, Lyon. 1997. p. 181

[44] Ramharack R. US Patent 5,391,406. 1995

[45] Wei SQ, Bai YP, Shao L. A novel approach to graft acrylates onto commercial silicones for release film fabrications by two-step emulsion synthesis. European Polymer Journal. 2008;**44**:2728-2736

[46] Patel HJ, Patel MG, Patel AK, Patel KH, Patel RM. Synthesis, characterization and antimicrobial activity of important heterocyclic acrylic copolymers. Express Polymer Letters. 2008;**2**:727-734

[47] Jeusette M, Peeterbroeck S, Simal F, Cossement D, Roose P, Leclere P, et al. Microscopic morphology of blends between a new "all-acrylate" radial block copolymer and a rosin ester resin for pressure sensitive adhesives. European Polymer Journal. 2008;**44**:3931-3940

[48] Xu GZH, Dong JP, Zhang JG, Severtson SJ, Houtman CJ, Gwin LE. Characterizing the distribution of nonylphenol ethoxylate surfactants in water-based pressure-sensitive adhesive films using atomic-force and confocal Raman. The Journal of Physical Chemistry B. 2008;**112**:11907-11914

[49] Taghizadeh SM, Mirzadeh H, Barikani M, Yousefi M. The effect of poly (vinylpyrrolidone) concentration on peel strength of acrylic/PVP pressure sensitive adhesive blends. Iranian Polymer Journal. 2007;**16**:279-285

[50] Bai Y-P, Zhao L, Shao L. Hybrid emulsifiers enhancing polymerization stabilities and properties of pressure sensitive adhesives. Journal of Applied Polymer Science. 2009;**115**:1125-1130

[51] Khanjani J, Pazokifard S, Zohuriaan-Mehr MJ. Improving dirt pickup resistance in waterborne coatings using latex blends of acrylic/PDMS polymers. Progress in Organic Coatings. 2017;**102**:151-166

[52] Khanjani J, Zohuriaan-Mehr MJ, Pazokifard SH. Microemulsion and macroemulsion polymerization of octamethylcyclotetrasiloxane: A comparative study. Phosphorus, sulfur, and silicon and the related elements. 2017;**8**(192):967-976. DOI: 10.1080/10426507.2017.1315418

Chapter 7

Liquid Thread Locking Solution for Machinery Assembly Industry

Chulsoo Woo

Abstract

Anaerobic liquid thread locking adhesives perform where mechanical fasteners fail. Just as snow tires grip the snowy track, liquid thread locking adhesives help fastener threads grip reliably. Based on anaerobic technology, these machinery adhesives do what mechanical methods had been designed to do but could not fulfill; they completely fill up the gap between the interfacing threads. Bolts secured with liquid thread locking adhesives retain clamping force better than mechanical fasteners not only reliably but also economically. A special nut or washer can cost up to five times as much as a liquid thread locking adhesive that reduces the cost of inventory, too.

Keywords: thread locking, mechanical assembly, anaerobic adhesive, fasteners, mechanical locking devices, tension scatter, K value, lubrication, break loose torque, bolt tension scatter, torque augmentation

1. Introduction

Mechanical assembly method had been used for a long time along the history of mankind. In 300 BC, the idea of Archimedes screw had been introduced to move fluid from lower level to upper level. This was the first application of using a screw. In 1500, spindle press had been introduced with a threaded assembly. In 1900, various mechanical locking devices had been introduced to assemble mechanical parts in many industries. In 1956, a liquid anaerobic thread locking adhesive had been developed and introduced to the machinery assembly industry. Since then, this one drop of chemical liquid has provided those R&D, production and quality engineers with an insight to eliminate those problems and issues of current mechanical fastener design they have strived to solve or improve for a long time. Various new mechanical fasteners have been designed and introduced continuously to industry over time with a high cost but still had an issue of ultimate loosening any time beyond mechanical assembly manufacturers' control.

2. Function of bolt and nut

When asked about the function of bolt and nut, almost everyone says that it is designed for assembly which is true. However, there is one more function which may be much more important than "assembly." Another function we must understand more is that it was designed for disassembly anytime when necessary, too. It means that the assembly should stay assembled no matter how long it is, but at the same time, it should be able to be disassembled by tool anytime. Does this sound logical to meet both requirements at the same time to you? In reality, design engineers have spent a lot of time to design mechanical fasteners that can achieve those two functions. Consequently, various mechanical fasteners have been introduced, and they will be reviewed more in detail on each fastener in this article.

In order for fasteners to do their job ideally, they should resist to external vibrations, shock, and impact, withstand shock and impact, tolerate differences in thermal expansions, protect against corrosion, and provide a consistent torque control during assembly.

3. Structure of bolt and nut assembly

Bolt is not a bolt; it is a spring. When tightening a bolt in the assembly, it means tensioning or slackening a spring. Bolt is pulling the flanges together as a "bolt spring." "Bolt spring" force must pull the flanges together more than the forces acting to push them apart. If "bolt spring" is too loose, the pressure stretches the bolt, and the flange opens and leaks. Therefore, "bolt spring" should prevent the flanges from separating; therefore, bolts are preloaded (stretched).

Clamp load is the force that holds a joint together. In a threaded assembly, the clamp load is equal in magnitude to the tension of the bolt but opposite in direction. Bolt tension is achieved after some friction loss at bolt head and thread. This clamp load is very critical as a bolted joint is designed so that the external forces never overcome the clamp load, and this is the major reason fasteners are used (see **Figure 1**) [1].

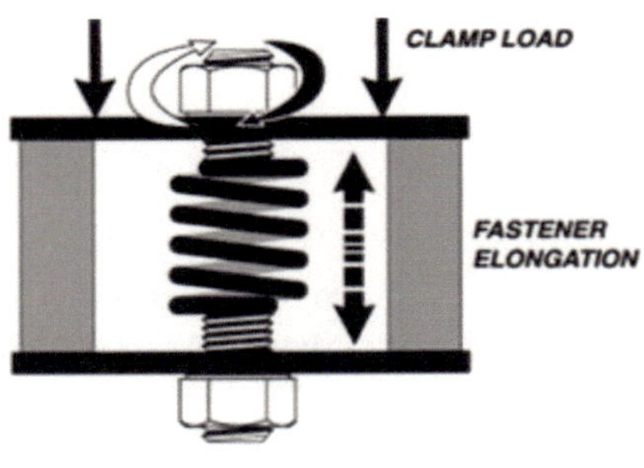

Figure 1.
Forces acting on a "bolt spring."

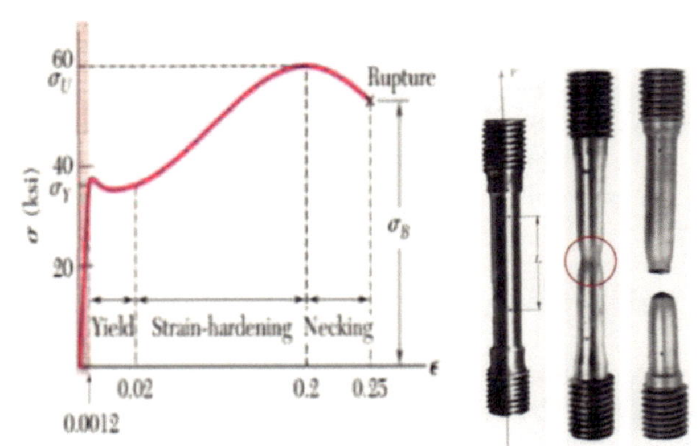

Figure 2.
Typical stress-strain curve of steel.

Bolt tightening torque stretches bolt shank to at least 65% of its yield strength. Yield strength is the stress at which the bolt shank starts to stretch. If the stress is beyond this point, then the bolt shank will break. As shown in **Figure 2**, tightening torque elongates the bolt. Being an elastic material, elongated bolt wants to retract. That is when a bolt tension is generated and the spring force clamps parts together. Red marked area shows the max. clamp load. Typically, the acceptable clamp load is in the range of 50–75% of bolt proof load (see **Figure 2**) [1].

4. Well-accepted myth

For such a long history of mechanical fasteners, those people related to designing, assembly, and repair have built some mindset on their own for a long time, it has become a conviction, then belief, and then it has become a myth to them. Here in this chapter, some of their "truth" will be shared to see if they are true.

- They think a properly tightened bolt according to the standard tightening guideline will never loosen under any circumstances. It is not true as when the friction in the thread and under the head of the bolt starts to drop, it does not take a long time for a bolt to loosen.

- They think it takes thousands of hours of vibration to loosen a bolt. It is not true as after side sliding starts, as few as 100 cycles are needed to loosen a bolt.

- They think fasteners take more torque to loosen than to tighten. It is not true as it is easier to go downhill than uphill.

- They agree with "I know it is tightened properly because I torqued it myself." It is not true as it does not matter whoever tightens.

5. Conventional thread locking fasteners

There are various conventional fasteners available in the market, in this article, eight different frequently used fasteners are explained on how each one works and what is the area of improvement regarding performance and cost.

5.1 Spring washer

Spring washer creates mechanical force against the nut and substrate surface which means mechanically "pushes the surface." What does this mechanical etching of the surfaces or how does this affect areas prone to corrosion? It may maintain the bolt tension in the beginning of assembly, but as time goes by, it loses its spring form and becomes flat due to the empty space between threads, and then, the bolt tension is reduced, and the assembly pressure is reduced, too, which results in the loosening of bolt and nut, and additional cost of spring washer is required (see **Figure 3**) [1].

5.2 Star washer

The star washer uses friction of the raised metal portions to mechanically "grip" the surface. This method allows an initial loosening as the washer rotates and grips into the substrate, and then bolt tension is already reduced, but it can prevent a further

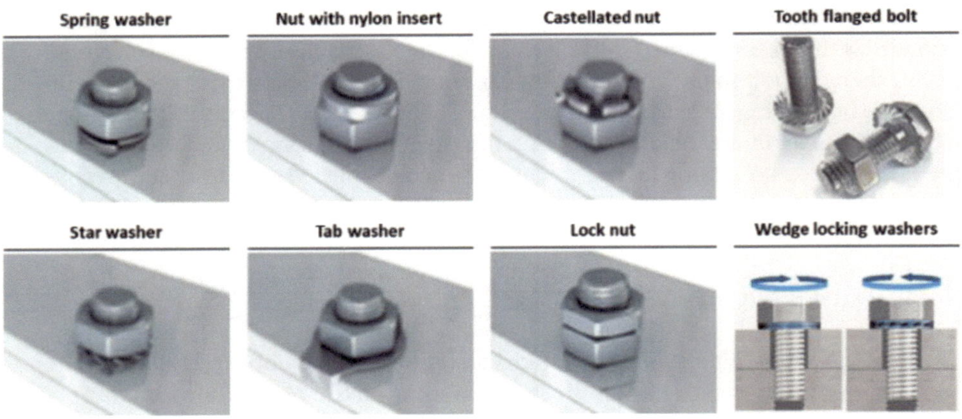

Figure 3.
Conventional thread locking fasteners.

loosening. This also does not seal, can damage the surface of softer or coated metals, and is not reusable, and additional cost of star washer is required (see **Figure 3**) [1].

5.3 Nut with nylon insert

Nylon or "plastic" deforms into void in the thread. This method allows an initial loosening, and the bolt tension is already reduced, but it can prevent a further loosening, and it is also recommended to be replaced after each disassembly with new nut in order to maintain optimal performance as when reused it loses a significant amount of its ability to fill the voids. Cost of nylon insert must be considered, too. Nylon also insulates the heat developed from thread friction making the assembly more susceptible to galling during installation. It is difficult to tighten because of nylon insert friction, too (see **Figure 3**) [1].

5.4 Tab washer

Tab washer mechanically overlaps the edge of a surface or flange. It acts as a normal washer, spreading the clamp load over a wide area. This method allows an initial loosening of bolt tension due to the empty space between threads. But due to its structure, it prevents a further loosening, and it does not seal or technically lock the assembly, and additional cost of tab washer is required (see **Figure 3**) [1].

5.5 Castellated nut

Castellated nut uses tension to mechanically "grip" the bolt or threads. Cotter pin is also sometimes used in conjunction with this device to prevent nut from falling off. However, cotter pin will not prevent loss of clamp load after initial loosening. It fails when vibrational load or impact causes the fastener to lose clamp load. This device does not seal the threaded assembly, and the high cost of castellated nut is required, too (see **Figure 3**) [1].

5.6 Lock nut

By "doubling nut" this method uses two nuts applied to the threaded fastener in order to provide a secondary force or lock to the initial nut. This does not significantly increase clamp load and may add additional stress to the threaded fastener.

This does not provide a seal and is prone to coming loose under vibrational loads and doubles the cost of the locking device (see **Figure 3**) [1].

5.7 Tooth flanged bolt

The "tooth" on the bolt grips and digs into the surface of the material or metal that is being assembled. This method also allows an initial loss of bolt tension as the tooth bolt rotates and digs into the bearing surface and stops. The tooth may dig into coated metals or damage the coating finish. This device does not seal the assembly. Under the vibration and impact, the bolt rotates until the tooth digs into the metal surface and finally grips the metal to prevent a further loosening; however, until then, pretty much amount of bolt tension is already lost, and a high cost of tooth flanged bolt is required, too (see **Figure 3**) [1].

5.8 Wedge-locking washers

A pair of wedge-locking washers with radial teeth on opposite surfaces grip the surface and are under the bolt head. It may damage coated or plated surfaces. It does not seal the joint and requires more torque to obtain target clamp load. This is the most expensive locking mechanism in the market (see **Figure 3**) [1].

6. Shortcomings of mechanical locking devices

As described on the above chapter, conventional mechanical locking devices have those shortcomings such as they lose under vibration, thermal expansion, and/or improper torque, and they do not seal threads, and they require extensive inventory of several shapes and sizes, and they are prone to rust and surface damage, and they have a wide tension scatter, and their torque is not well augmented, and they are more costly than liquid thread locking adhesive.

6.1 How does bolt and nut loosen?

Threaded fasteners were designed to be removed anytime. The problem is that it loosens unintentionally. Thread is a rolled cone and unrolled thread is an inclined plane. As shown in **Figure 4**, "F sin a" of the induced load assists the loosening. When an object is in an equilibrium state between gravity and friction on an inclined surface, and then, if there is a sideway transverse impact or vibration applied to the

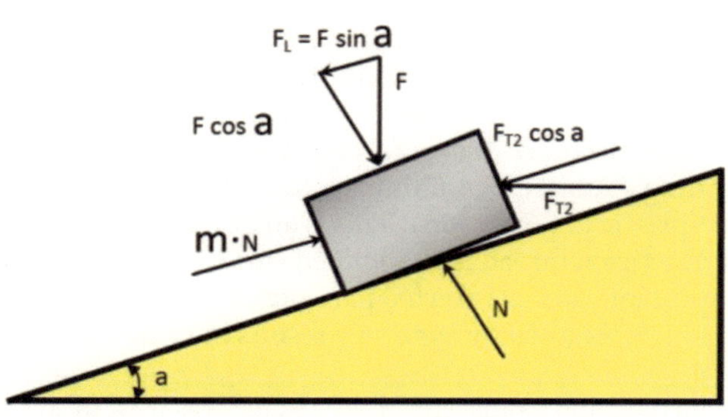

Figure 4.
Object on an inclined plane.

object, the object easily slides down meaning loosening of the fastener also downhill is 30% less effort (easier to slide down) than uphill (going up) [1, 2].

6.2 Reasons for failure

6.2.1 Relaxation

In a flanged assembly, a cut gasket receives a high pressure from bolt tightening. In the beginning, cut gasket has a certain level of flexibility to push back the flange pressure to achieve a sealing. As the flange pressure continues to press the cut gasket, the cut gasket comes to a point that it no longer pushes back and loses a flexibility. A gradual decrease of cut gasket thickness over time is called "creep." As the cut gasket starts to reduce its thickness due to the flange pressure, the cut gasket becomes "compression set." Once the compression set was done, the bolt tension of the flange is reduced, and then, the flange surface pressure reduces leading to a leak, and the bolt and nut assembled become susceptible to an unintentional loosening against external vibration and impact. The same is applied when the flange substrate is soft enough to be deformed by the bolt and nut with a high tightening torque. The bolt tension is reduced, and the assembly is easy to loosen. Another one is "settling" of the substrate surface as its surface irregularity. Especially, peak is flattened by the flange pressure to settle down a bit. However, this settling is very minimal to cause the reduction of bolt tension (see **Figure 8**).

In a flanged assembly, initial settling of flange and creeping of cut gasket might be compensated by the bolt elasticity to provide a sealing and no bolt loosening, but as time goes by and the cut gasket thickness becomes thinner, then bolt elasticity can no longer compensate the loss of flange pressure. The bolt tension will be reduced, and bolt and nut will loosen eventually. Therefore, it is a routine work and guideline for mechanical assembly manufacturers that they do a "re-torqueing" of the bolt and nut assembly before shipping after they manufacture and store at the warehouse. Again, such above method will add additional cost to the assembly [1, 2].

6.2.2 Loosening

After the tightening procedure, the clamp load is maintained by the preload in the bolt. This occurs because the bolt has been stretched like a spring, and the tension acts to pull the flange or substrate. As shown in **Figure 5**, an object is in an equilibrium state (no move) in an inclined surface as the gravity and friction are applied equally. A test was done using a simulated object and inclined plane to see how the object moves when a transverse force is applied to the side of the object. When it was pushed sideway, it slides down and then stops. Then another sideway push, it slides down and stops again. Consequently, when there was a continuous sideway pushes, the object continued to slide down. As bolt and nut are tightened more and more, it means that one thread climbs up the mating thread. It also means that the height energy becomes higher and has a high tendency to slide down when there was a loosening parameter combining external ones (impact and vibration) and internal one (plastic deformation). This is an exact representation of a real threaded assembly as each thread is an inclined surface (see **Figure 5**) [1].

When there were internal or external parameters, such as impact and vibration, it generates a sliding force and consequently results in insufficient clamp load and loosening.

A special testing machine (transverse shock tester) was used to compare how fast the mechanical locking devices lose the initial bolt tension under shock and vibration. Bolt and nut are used together with locking devices such as spring washer, nylon insert, and

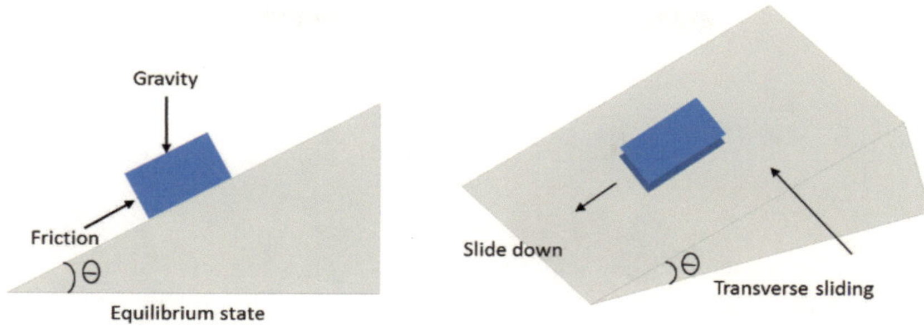

Figure 5.
Transverse loosening of a bolt and nut.

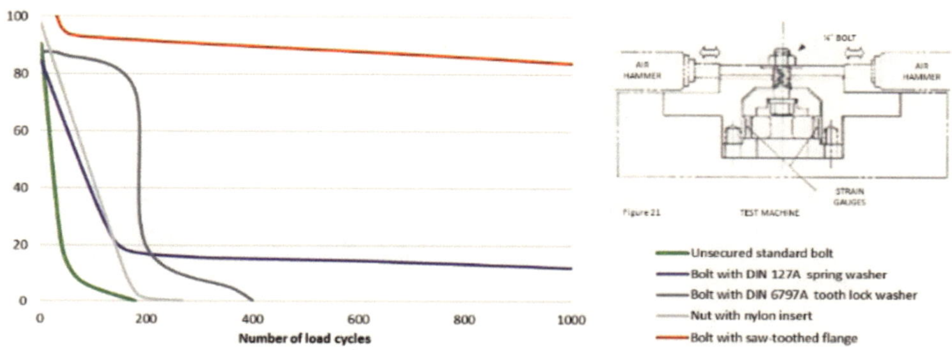

Figure 6.
Transverse shock test.

serrated bolt. They are assembled with a pre-determined tightening torque in the testing machine. Then, air hammer is actuated by hitting both sides of the assembled part. The machines monitor the loss of initial bolt tension as time goes by. Those mechanical fasteners assembled lose their initial bolt tension within "20 s" (see **Figure 6**) [1, 2].

6.2.3 Differential thermal expansion

When substrates have a different thermal expansion coefficient, it causes a sliding movement when subject to high or low temperature. It results in loss in bolt tension, reduction in clamp force, and consequent assembly failure (see **Figure 8**).

6.2.4 Corrosion

When mechanical locking devices are used for assembly, metal-to-metal contact area is 15% between male and female threads only and between bolt head and substrate. The rest 85% is all empty which is susceptible to the ingression of moisture or other chemical environments resulting in the corrosion inside. Once it is corroded, it is quite difficult to disassemble by tools. Then they have to end up with cutting the whole fastener assembly by torch and procure new fasteners again. For bigger size bolt, quite often, the assembly failure occurs (see **Figures 7** and **8**) [1].

6.2.5 Tension scatter

The flange assembly tightened with fasteners often has a flange sealing failure. It is difficult to achieve a uniform, consistent flange pressure with normal bolt and

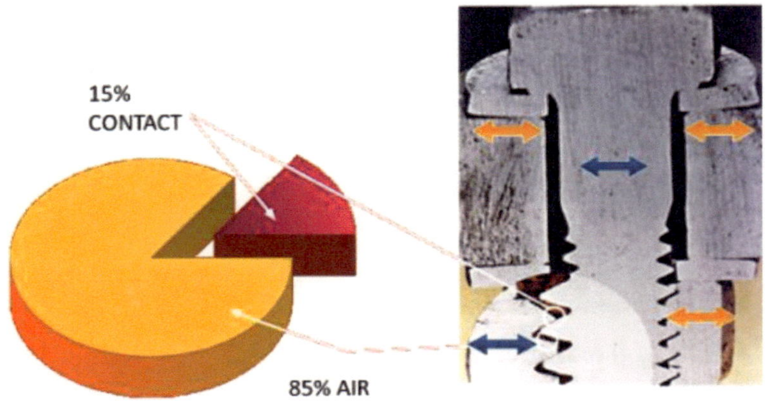

Figure 7.
Empty space in a threaded assembly.

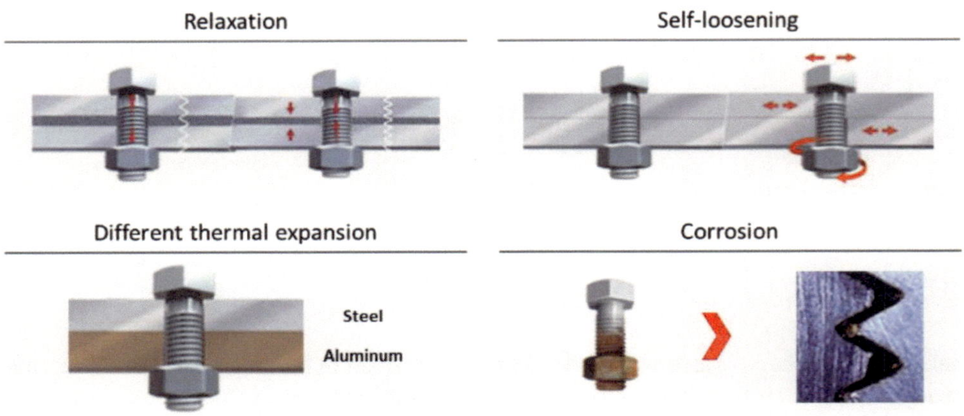

Figure 8.
Reason for assembly failures.

nut fasteners because each bolt generates a different bolt tension even though it was tightened equally according to the standard tightening guideline. The same size bolts are made by the same supplier and same spec. All end up having a different bolt tension after tightening. Then the flange pressure becomes high at one side of bolt assembly, and the flange pressure becomes low at other sides of bolt assembly, and then, one side has higher stress, and the other side has lower stress. It is suppose that five bolts were used to assemble a flange assembly. The same tightening torque was used to tighten each bolt. But a different bolt tension (or flange surface pressure) is achieved for each bolt because each bolt has a different surface roughness on the threads. This different surface roughness on the thread surface generates a different remaining bolt tension after some portion is consumed by the friction on the thread, and consequently, through the flange sealing area with a low flange surface pressure, the leak occurs. This can cause a long-term reliability issue of sealing, and it must be addressed properly to prevent such failure (see **Figure 9**) [1].

6.2.6 Torque augmentation

When the mechanical assembly is done with fasteners on a flange, it is expected that the assembled fastener should maintain in the assembly the 100% of the tightening

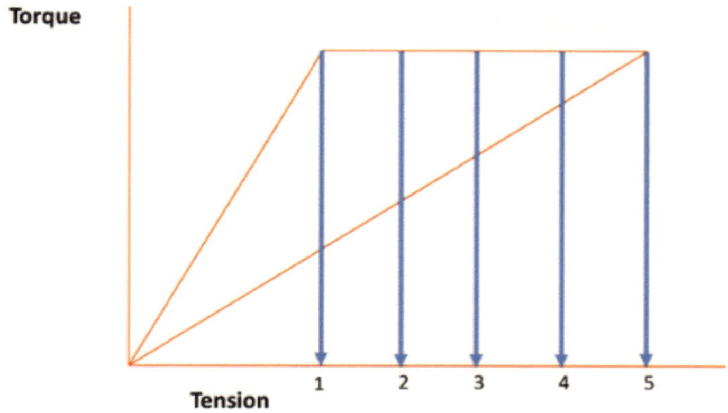

Figure 9.
Wide tension scatter.

torque applied according to the standard guideline. However, in reality, it is not. When "10 N m" tightening torque was applied to tighten fasteners, it takes around "7 N m" to loosen the assembly. It means that around 30% of tightening torque is lost during the assembly to overcome the friction generated between male and female threads and between bolt head and substrates. This explains that the assembly maintains only 70% of the tightening torque. Therefore, when there were parameters to loosen the fastener, it would have a high possibility to be susceptible to such loosening factors and finally loosen. A proper measure should be addressed to prevent this failure. At the same time, it is important to understand that the torque augmentation becomes thin as the bolt size, and its tightening torque becomes bigger because the gap between threads is getting bigger, and it is not an ideal condition for anaerobic adhesive to polymerize and lock the thread in place. As shown in **Figure 10**, the lines of torque augmentation for bolt and nut assembly of "without anaerobic adhesive" and "with adhesive" are not parallel; they will meet at some point. In the situation of big bolt and nut, thread locking is not a main point as a very high tightening torque is applied. When the break loose torque for bolt and nut assembly was measured with and without anaerobic adhesive, the result is very identical meaning there is not much torque augmentation due to too high tightening torque and big gap between threads. Therefore, thread "seizing" or "galling" is more important point which they have to use an extreme high-temperature and high-pressure lubricant product to prevent the thread seizing and galling (see **Figure 10**) [1].

When a bolt and nut were tightened with a tightening torque, it has to overcome the friction generated between threads and bolt head or nut to part the surface. As shown in **Table 1**, a comparison test has been done to see the difference between "Unified National Coarse" (UNC) thread and "Unified National Fine" (UNF) thread. Coarse thread has more bolt tension than fine thread, and fine thread has more thread and head friction than coarse thread. But on the other hand, coarse thread has more torque loss than fine thread. The distribution of tightening torque is shown in **Table 1** [2].

6.2.7 Galling

Galling is a wear caused by friction and the buildup of heat between two sliding thread surfaces. When a relatively big size bolt and nut are assembled or loosened for repair, a very high heat is generated; this heat fuses and melts the two sliding thread surfaces and causes the galling. The same issue occurs when a low thermal conductivity of metal is used as a fastener for assembly or disassembly. A care must be taken when stainless steel fasteners are used as they have the lowest thermal conductivity. **Figure 11** shows thermal conductivity of major fastener metals [1].

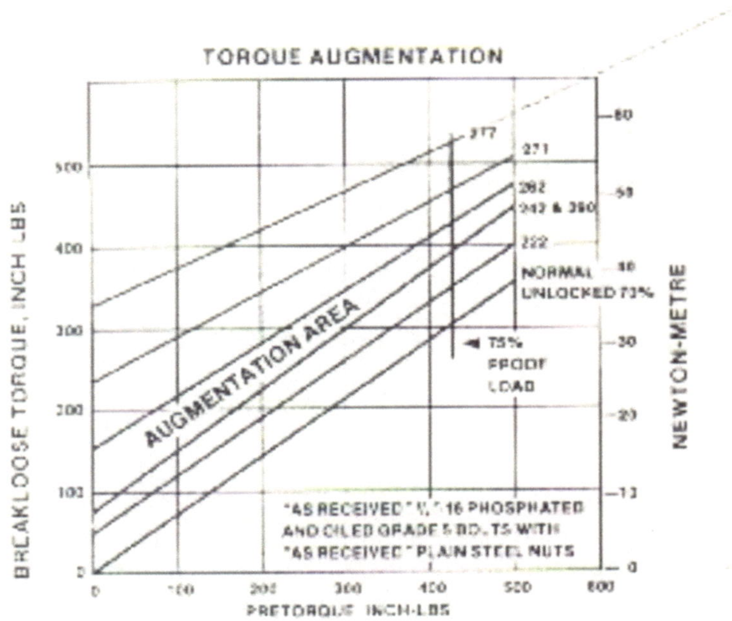

Figure 10.
Torque augmentation.

	UNC	UNF
Bolt tension	15%	10%
Thread friction	39%	42%
Head friction	46%	48%
Tightening torque	100%	100%
Loosening torque	70%	80%

Table 1.
Distribution of tightening torque.

Copper	231	**Least likely to gall**
Aluminum	136	
Brass	69	
Zinc	67	
Steel	32	
Stainless Steel	8	**Most likely to gall**

Figure 11.
Thermal conductivity of metals (Btu/hr-ft-F).

6.2.8 Cost

Those conventional fasteners require additional items such as flat washer, spring washer, star washer, and tab washer, and some of them were designed especially such as nut with nylon insert, castellated nut, and tooth flanged bolt.

They are much more expensive than using a plain bot and nut. It would be ideal to have a way to lock the normal bolt and nut in a way we want but with a minimal cost addition.

7. Liquid thread locking adhesive

This chapter explains everything about anaerobic liquid thread locking adhesive such as its curing mechanism and benefits that can help assembly industries. This is a liquid single-component product that fills all voids to unitize the bolt and nut, and it cures to a thermoset plastic and creates material closure. It also maintains friction by adhesion, resists a vibration, and prevents bolt and nut from loosening.

7.1 Anaerobic curing mechanism

When adhesive is applied onto one side of metal substrate, then adhesives contact with metal ion (Fe++, CU++) for some anaerobic initiator to be broken into free radicals which are unstable, but the presence of air to contact adhesive is stronger; therefore, it remains as a liquid. As a next step, when the other mating part comes in contact with the first part with adhesive in between, then more anaerobic initiators contact more metal ions to be broken into more free radicals, and at the same time, oxygen contact is blocked by both substrates, and then, those unstable free radicals react with a monomer nearby which becomes unstable, too, and then, it reacts with other monomers nearby, and gradually, all the monomers react and become polymerized (cure) (see **Figure 12**) [3, 4].

In case of inactive metal substrates that contain low metal ions such as Cu++ or Fe++ and have a slow reaction with photoinitiator and slow cure, an acceleration method such as primer or heating can be used to speed up the curing. Anaerobic adhesives become a very reliable thermoset plastic after cure [2].

7.2 Liquid thread lockers and benefits

The gap in the threaded fasteners after assembly has always been the issue ever since it has been developed and used. It has led various unwanted situations that had to be corrected and improved with additional cost and time. Liquid thread locking adhesive is one component that fills all the inner space of assembled threaded fasteners and cures to a tough thermoset plastic. It is called as "Conquest of Inner space." It is surprising to know that the inner space in the threaded assembly has had issues. The liquid thread locking adhesive solves various issues that mechanical fasteners

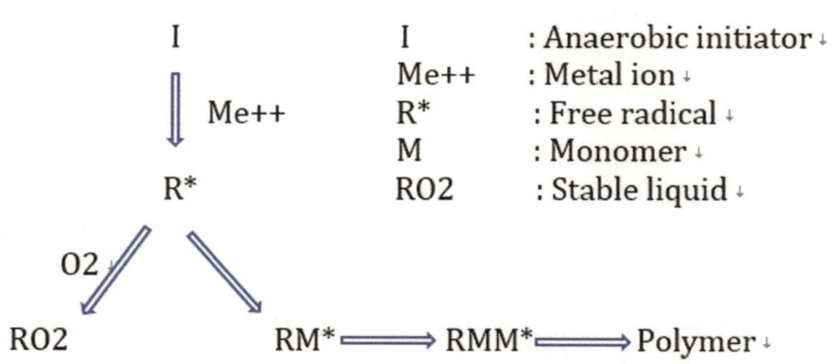

Figure 12.
Anaerobic cure mechanism.

used to have. It is the best choice that provides a cost-saving and reliability improvement of a final mechanical assembly and really conquests the inner space.

The liquid thread locking adhesives have various viscosity ranges from just like water to a paste. Water-like viscosity is used for assembly that has to be assembled first due to structural or locational reason, and then, this product can be applied to wick into the inner space by capillary action and lock the assembly in place. The extremely high temperature requires a pasty type thread locking adhesive that contains a special filler to resist such high temperature. It also has a different strength depending on the application condition and requirements. A repairable type can be used for easy repair as it can be disassembled by a simple tool any time, whereas a permanent strength type can be used for the application that they do not need to repair. Its thixotropic property allows a controlled dispensing that helps prevent it from flowing to adjacent area. Lubricity is another benefit that allows easier assembly and reduces a tension scatter for more reliable assembly. As it contains a fluorescence, it becomes very useful when checking its presence, quality control, and troubleshooting. It also has received various approvals such as drinking water approval, food grade approval, military approval, etc. This liquid thread locking adhesive can be applied with various dispensing equipment ranging from manual volumetric hand pump, semiauto dispenser, and fully automatic dispenser for automation, too.

7.2.1 Maintain clamp load

Designers are well aware of designing for fatigue resistance, and a properly clamped joint is designed. The immediate question is what should be the design load of the joint? We know that a clamped joint tighter than applied loads will not experience cyclic loads on the bolts. However too high clamping force will affect the performance of bolt as the yield point of bolt has to be considered, and it also can damage the assembly part to cause a plastic deformation and consequent loss of bolt tension and loosening. Those specially designed fasteners such as serrated bolt will loosen under impact or vibration until the tooth locks into the part surface, and ultimate bolt tension is reduced. Another one is spring washer. Even though it can achieve an initial clamping force, the push back force of spring washer is reduced as time goes by, and it leads to a reduction of bolt tension and consequent loosening. Liquid thread locking adhesives are the most effective in maintaining clamp load without bolt failure and eliminating self-loosening tendencies [2].

7.2.2 Prevent loosening

The best way to evaluate a thread locked connection is to test its behavior under load cycles in a dynamic test machine. The lower the loss in bolt tension, the more reliable the assembly. The bolt test machine for comparative test of thread locking system is called TSV (transverse shock and vibration) machine. Pneumatic powered hammers force relative movement of the stressed parts by a displacement. At the same time, the existing bolt tension is continuously measured for the number of load cycles. Characteristic clamp load retention curves of various locking methods can be compared with the results. To determine clamp load retention curves for various locking devices, a bolt test machine like the Junker machine was used (see **Figure 6**). A bolt and nut are assembled with two metal parts that are situated between two air hammers on both sides. After the nut was tightened to a specified tightening torque, a bolt tension is achieved. Two air hammers are actuated by hitting the assembled part at 600 rpm continuously. This machine is connected to a computer monitor that shows an initial bolt tension and its change as time goes by. Typical fasteners like flat washer, spring washer, star washer, etc. quickly lose initial bolt tension as air hammers hit the assembled part. Initial bolt tension loss means a sliding of thread due to the space inside and slope shape of thread. There is one, serrated bolt head that maintains an initial bolt tension even though it shows a drop of bolt tension at the initial air hammer impact. However, the most important thing is that the initial bolt tension should maintain constantly as initial bolt tension loss means already a failure of the threaded bolt assembly. Liquid thread locking adhesives fill the gap between threads and lock the bolt and nut in place, and even though there was an external air hammer impact, the assembly maintains an initial bolt tension constantly. As shown in **Figure 6**, mechanical fasteners lose initial bolt tension drastically in the beginning and leave almost no tension in "20 s," whereas liquid thread locking adhesive maintains a flat line of bolt tension as time goes by [2].

7.2.3 Break loose strength

The structure of assembled mechanical fastener has an empty space inside. Once it starts to loosen, it loses to a very low bolt tension as there is no resistance to overcome the loosening force, and it only has 70–80% of tightening torque as the rest was used to overcome the friction. For liquid thread locking adhesive, break loose torque may be used as a measure for wetting, adhesion, and degree of curing. In general, there is no direct relation between break loose torque and self-loosening resistance. In practice, it is often required that threaded assemblies be capable of disassembly using normal tools. Therefore, products with low and high break loose strength have been introduced to satisfy those application requirements. For better clarification, there are two different types of break loose strength. One is medium strength (repairable type) that can be disassembled with a simple tool, and the other is high strength (permanent type) that is difficult to disassemble by a tool. Care should be taken when disassembling the parts locked with a permanent type adhesive as it may damage the bot head and shank into failure while trying to disassemble. More than 250°C localized heating is required to soften the cured adhesive and disassemble the assembly with tool while hot [2].

The break loose torque strength can be decided by different parameters as medium strength adhesive can achieve a high strength when used to a big bolt and nut. Therefore, care should be taken when considering a right thread locking adhesive. Those parameters can be a bolt size as mentioned on the above as the overall bonding area becomes bigger, and other parameters are material of bolt and nut as it may have a different strength according to active metal and inactive metal substrates.

One of the benefits of liquid thread locking adhesives is that it can be used to various sizes of fasteners regardless of bolt size and diameters. While conventional special locking fasteners require large inventories, thread locking adhesives can be used on all standard and special bolts. Searching for or ordering special locking fasteners is eliminated [2].

7.2.4 Differential thermal expansion

If a liquid thread locking adhesive was used and cured in between threads of male and female assemblies, in spite of differential thermal expansion, it prevents the sliding movements of the assembly from occurring due to 100% filling inside; therefore, it maintains a consistent and stable clamp load (see **Figure 8**) [5, 6].

7.2.5 Corrosion

A liquid thread locking adhesive when applied fills all the inner spaces and cures to a thermoset plastic; therefore, even under a very severe environment condition, no ingression is allowed inside, and it allows an easy disassembly when necessary (see **Figure 7**) [5, 6].

7.2.6 Tension scatter

A liquid thread locking adhesive has a lubrication during bolt and nut assembly which allows the bolt tension of each tightened bolt to be very close which means the tension scatter becomes very narrow, and consequently, each bolt tension will end up being very consistent. This way the whole assembly will have a consistent flange pressure and increase the leak free reliability, value, and brand image of the assembly (see **Figure 13**) [5, 6].

In order to understand lubrication K value, an equation is used to describe the relationship between torque and tension:

$$T = K \times F \times D \tag{1}$$

The K value or torque coefficient is an empirical constant value which takes into account friction and the variable diameter under the head and in the threads where friction is acting. It is determined experimentally to describe the relationship between torque and tension.

where

T = torque (N m).

K = torque factor.

F = clamp load (N).

D = nominal diameter (mm).

Eq. (1) torque and tension relationship.

Values of K can be determined experimentally. The lower the K value, the better the lubricity. The lower the K value, the poorer the lubricity. As shown in **Table 2**, liquid thread locking adhesive has better lubrication than lubrication oil on most of materials (see **Table 2**) [2, 5, 6].

Fastener threads without lubricating oil, due to different surface roughness, yield a wide bolt tension scatter, and it leads to an unstable threaded assembly on loosening or leak. As shown in **Table 2**, lubrication oil shows a range of K value of 0.13–0.22, and liquid thread locking adhesives has a K value in the range of 0.09–0.24. It means it has lower K value than lubrication oil for many materials, and using liquid thread locker means it is as if using a lubrication oil together. It is another benefit and liquid thread locking adhesive has lower K value; therefore, it can yield controlled, consistent torque tension curves and narrow tension scatter and therefore is excellent for a high-quality in-line production. Factors that influence K value are fastener and flange materials, insertion speed, bolt quality, adhesive product selected, thread tolerance, and surface finish [2, 5].

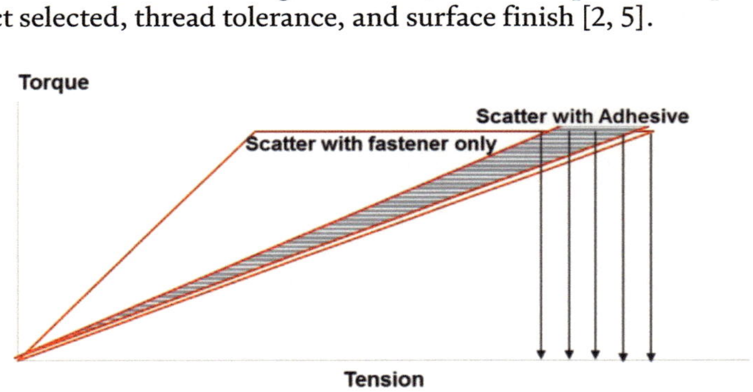

Figure 13.
Narrow tension scatter with liquid thread locking adhesive.

Substrate	Oil	Thread locker
Steel	0.15	0.14
Phosphate	0.13	0.11
Cadmium	0.14	0.13
Stainless 404	0.22	0.17
Zinc	0.18	0.16
Brass	0.16	0.09
Silicone bronze	0.18	0.24

Table 2.
Typical K values—Lubricating oil and thread lockers on materials.

7.2.7 Torque augmentation

A liquid thread locking adhesive provides a torque augmentation when the bolt and nut assembly with adhesive was disassembled higher than the tightened torque. As shown in **Figure 10**, the bolt and nut assembly without adhesive showed a lower break loose torque than tightening torque which is 70% only. However, the break loose torque with adhesive showed a higher figure than tightening torque (more than 100%) which can make the assembly more stable, reliable, and ensure a longer life (see **Figure 10**) [1].

7.2.8 Galling

When using a liquid thread locking adhesive during assembly, it provides a lubrication that reduces the friction between threads, heat buildup, and, consequently, galling.

7.2.9 Thread locker lock and seal

As mentioned earlier, mechanical fastener assembly is susceptible to a loosening, leak, and corrosion inside. Thread is in an inclined plane and easy to slide down (loosen) with an external transverse impact or vibration. It also leaves an inner space even though how high it was tightened, and it does not provide a sealing against fluids. Such inner space also allows external environment to penetrate and cause a corrosion that makes the easy disassemble very difficult. In many cases, torch has to be used to cut the threaded assembly and use a new one to reassemble. Liquid thread locking adhesives provide a locking and sealing as it fills all the inner spaces to cure and lock it in place against external impact and vibration. It also provides a sealing against fluid such as hydraulic system thread sealing as the cured thermoset plastic has a resistance against various fluids and chemicals. It also protects inner surface against corrosion; therefore, disassembly becomes very easy anytime. So, the life of mechanical assembly and assembly quality image are improved and increased. This also allows easier and cost-effective bolt-hole design as a cured thermoset plastic has a resistance against various fluids.

7.2.10 Cost efficiency

Liquid thread locking adhesives cut costs by replacing expensive special locking bolts and nuts. Less costly standard bolt or nuts can be used. **Table 3** compares the costs for the "ribbed flange bolt" and "adhesive" locking methods. The costs of normal shaft bolt secured with liquid thread locking adhesive are defined as 100%. The chart shows the relative increases or reductions in cost with surface-compacting

Method	Material costs	Assembly costs	Total costs
Ribbed flange bolt	130%	65%	118%
Liquid thread locker	100%	100%	100%

Table 3.
Cost comparison between "ribbed flange bolt" and "adhesive" locking method.

ribbed flange bolts. The parameter that makes the "ribbed flange bolt" expensive in overall cost is because of its material cost. Over the long years, mechanical designers have tried to find a way to keep the assembly safe against loosening without control, and they have come up with those expensive fasteners, but they could not prevent a complete "loosening prevention" either [2, 5, 6].

As shown in **Table 3**, liquid thread locking adhesive is a less costly method. The cost advantages coming from the simultaneous sealing effect and the use of through holes instead of blind holes have also not been considered. The liquid thread locking can be provided right from the design stage, reducing the costs of overall structure design and providing a guaranteed lock against loosening [2, 5, 6].

7.3 Adhesive application

Liquid thread locking adhesive can be applied to bolt thread in most of the cases to show an effective locking of threads. But for some critical or special design applications, liquid thread locking adhesives can be applied to both male and female threads to ensure a complete filling of the inner space and to cover an effective thread engagement length. The important factor is that the inner space should be filled completely. All "50-ml" and "250-ml" bottles of liquid thread locking adhesives are equipped with dispensing nozzles for use directly from the bottles. The flexibility of this method can be used to variations in the quantity of adhesive dispensed from one application to another. Dispensing systems allow accurate control of quantity. Liquid thread locking adhesive dispensing equipment has been designed for the specific properties of its adhesives to ensure trouble-free, reliable dispensing. Various dispensing equipment are available according to application requirements. For simple manual dispensing, volumetric hand pump that can be connected to the adhesive bottle directly can be used to dispense a controlled amount to bolt thread or threaded tap hole. For semiauto dispensing, a dispenser with a reservoir and controller can be used, and it can fit various applications in the different mechanical assembly industries. For automatic dispensing, a dispenser with reservoir, controller, and sensor that can be interlocked with a whole assembly line can be used. This automatic dispenser in many cases is used together with three axes or multiaxis robot for precise and accurate dispensing for fully automated production line [2].

7.4 Major products available

There are also various choices of adhesives according to each product performance as follows:

7.4.1 Go-to product

This is a first choice product for thread locking applications when a fast fixture time is required not only for active substrates and inactive substrates but also for some oily surface without the need of cleaning and when a high-temperature resistance of up to 180°C is required. This product has both medium strength and high strength.

7.4.2 Food grade product

This is a right choice for applications when thread locking of bolt and nut for manufacturing or repairing of equipment used for making food such as beverage, chips, bread, soup etc. It confirms to the requirement of FDA. This product has a medium strength that they can repair by hand tools easily when necessary anytime.

7.4.3 Ultra-high-temperature product

This is a product for application when then a threaded assembly must endure an ultra-high temperature environment up to "350°C." This product has both medium strength and high strength.

7.4.4 Environmentally green product

This is a product that carries no warning sign on the product label and material safety data sheet (MSDS) such as no R phrases associated with raw materials, no product hazard symbol, and no product risk phrases, in other words, no skin irritation, nontoxic, etc. This is simply an environmentally green product and the best choice when the assembly site requires a safe health and safety (H&S). This product has a medium strength and high strength.

7.4.5 Semisolid product

Semisolid grade is a compact stick design in a semisolid state just like a glue stick. It can be used mainly for repair application where worker can carry and apply easily to vertical or ceiling area of applications. This product has a medium strength and high strength.

7.5 Major terms

7.5.1 Breakaway torque

This is a strength of cured adhesive when bolt and nut were assembled by hand without on-torque. This measures a strength of adhesive only.

7.5.2 Prevailing-off torque

This is a strength of cured adhesive after breakaway torque was measured and the strength after 180° turn. This also measures a strength of adhesive only.

7.5.3 Break loose torque

This is a strength of cured adhesive when bolt and nut were assembled with an on-torque together with adhesive. Measuring of the strength is done the same as the breakaway torque.

7.6 How to identify application

It is critically important to understand the exact requirements of each application in order to provide a right solution. There are 10 key questions to ask to clarify and identify the application situation and requirements.

Application: this is the first question to clarify what is the configuration of the application such as thread locking, thread sealing, etc.

Material: depending on the material, products to recommend can vary as it will affect cure speed and strength such as mild steel, stainless steel, etc.

Dimension (size): depending on the size of application, a different bonding technology can be selected.

Bond line gap: each technology and its products have different gap filling capabilities.

How to assemble parts: depending on the shape of application and production cycle, products to recommend may change.

Fixture time: this will give us a choice of selecting a right product and/or right primer or heat depending on the requirement.

Need to maintenance: this can help select a right product as there are repairable strength product and permanent strength product.

Load: depending on a different type of load and its direction, a proper product can be chosen.

Environment: depending on the different chemical requirements, a proper product can be selected such as temperature, chemicals, etc.

Dispenser: those fittings and feedlines that come in contact with adhesive should be compatible in order not to affect product negatively.

8. Pre-applied thread locking adhesive

For some applications that they do not want a liquid chemical available in their workshop or their assembly line is automated with fasteners feeding so cannot use a dispensing but want to increase the reliability and reduce an assembly cost, pre-applied thread locking adhesive is available in the market. It contains microcapsules within the dry-to-touch coating. When the parts are assembled, the microcapsules will crush and release the active ingredient, and the adhesive will cure like a liquid anaerobic having a similar break loose strength and resistance to various chemicals. This adhesive coating has a designated color for each product that provides a guideline on quality check even before and after assembly. As this is a water-based product, it is environment friendly and no need to install ventilation or no need to receive authority's regular solvent smell inspection. It also has a good lubrication during assembly. Coating machines and types are available in the market depending on the quantity to be coated, and it also provides a long storage shelf life after coating [2].

9. Conclusions

Liquid anaerobic thread locking adhesive can allow the mechanical assembly industry to solve various issues that have been present for a long time whether it was recognized by them or not. It can save the overall assembly cost and increase the reliability of their assembly making their product more competitive in the market place. For performance wise, it not only locks against vibration, shock, and thermal cycling but also seals against corrosion and galling. Medium strength adhesive can be easily disassembled by simple tool. It also provides better clamp load retention than all mechanical locking devices. For cost-saving wise, it provides threaded assemblies that can reduce a costly downtime. "One size fits all" can be universally applicable for a wide range of fastener sizes. Its ease of automation reduces assembly costs and increases throughput. It has lower cost per unit than most conventional thread locking fasteners.

Author details

Chulsoo Woo
Henkel (China) Investment Co., Ltd, Shanghai, China

*Address all correspondence to: cs.woo@henkel.com

References

[1] Woo C, Robert D, Robert V, Pradhyumna I. Reliable and Cost-Effective Locking and Sealing of Fasteners. Shanghai, China: Henkel Corporation; 2008

[2] Alan L, Rudolf N. Loctite Worldwide Design Handbook. 2nd ed. Dublin Ireland: Henkel Corporation; 1998

[3] Woo C, Rudolf N, Robert D. Loctite Solution for Thread Locking Technology. Ontario, Canada: Henkel Corporation; 2017

[4] Woo C. Functional Adhesive Trend for Assembly Industry. Shanghai China: Henkel Corporation; 2019

[5] Woo C, Rudolf N, Robert D. Conventional Method of Thread Locking. Ontario, Canada: Henkel Corporation; 2017

[6] Woo C, Rudolf N, Robert D. Anaerobic Technology. Garching, Germany: Henkel Corporation; 2017